Incident Management in Intelligent Transportation Systems

For a complete listing of the *Artech House Telecommunications Library,*
turn to the back of this book.

Incident Management in Intelligent Transportation Systems

Kaan Ozbay
Pushkin Kachroo

Artech House
Boston • London

Library of Congress Cataloging-in-Publication Data
Ozbay, Kaan, 1964-
    Incident management in intelligent transportation systems / Kaan Ozbay, Pushkin Kachroo.
        p.   cm. — (Artech House intelligent transportation systems library)
    Includes bibliographical references and index.
    ISBN 0-89006-774-0 (alk. paper)
    1. Traffic congestion—United States—Prevention.   2. Electronic traffic controls—United States—Management.   3. Traffic flow—United States—Management.
    4. Intelligent Vehicle Highway Systems—United States.   I. Kachroo, Pushkin.
II. Title.   III. Series.
HE336.C64097   1999
388.3'142'0973—dc21   99-17751
  CIP

British Library Cataloguing in Publication Data
Ozbay, Kaan
    Incident management in intelligent transportation systems. —(Artech House intelligent transportation systems library)
    1. Traffic engineering   2. Information storage and retrieval systems—Traffic accidents
    3. Electronics in transportation
    I. Title   II. Kachroo, Pushkin
    388.3'12
    ISBN 0-89006-774-0

Cover and text design by Darrell Judd

© 1999 Artech House, Inc.
685 Canton Street
Norwood MA 02062

All rights reserved. Printed and bound in the United States of America. No part of this book may be reproduced or utilized in any form or by any means, electronic or mechanical, including photocopying, recording, or by any information storage and retrieval system, without permission in writing from the publisher.
    All terms mentioned in this book that are known to be trademarks or service marks have been appropriately capitalized. Artech House cannot attest to the accuracy of this information. Use of a term in this book should not be regarded as affecting the validity of any trademark or service mark.
    International Standard Book Number: 0-89006-774-0
    Library of Congress Catalog Card Number: 99-17751

10 9 8 7 6 5 4 3 2 1

Contents

	Preface	**xiii**
1	**Introduction**	**1**
1.1	Highway Congestion	1
1.2	Impact of Incidents on Highway Congestion	4
1.3	Incident Types and Impacts	6
1.4	Incident Management	9
1.5	Agencies Involved in Incident Management	12
1.6	Incident Management Process	14
1.7	Problems in Incident Management	15
2	**Review of Incident Management Systems**	**21**
2.1	Introduction	21
2.2	Proposed Implementation Frameworks for Incident Management Support Systems	22
2.2.1	System Requirements and Characteristics	23
2.2.2	Blackboard Architecture	28
2.3	Incident Management Frameworks Based on Expert Systems	31

2.4	Incident Management Systems Based on Geographical Information Systems	33
2.5	Summary	36
	Review Questions	37

3 Wide-Area Incident Management Support System Software — 41

3.1	Design Considerations	41
3.1.1	Overall Concept	42
3.1.2	Framework for Integration	43
3.2	Application Design	45
3.2.1	Decision Support Modules	46
3.2.2	Duration Estimation Module	47
3.2.3	Delay Calculation Module	48
3.2.4	Response Module	49
3.3	Software Implementation	50
3.3.1	Software Implementation Architecture	52
3.3.2	Application Development	54
3.3.3	Data-level Integration	56
3.3.4	Command-level Integration	56
3.4	Summary	57
	Review Questions	58

4 Incident Detection — 61

4.1	Introduction	61
4.2	What Is Incident Detection?	62
4.2.1	Traffic Surveillance and Data	62
4.2.2	Analysis of Traffic Data	62
4.2.3	Importance of Incident Detection Time	63
4.3	Effect of Incident Detection Time on Overall Incident Duration	64
4.4	Incident Detection Issues	66

4.4.1	Surveillance Issues	66
4.4.2	Algorithmic Issues	69
4.5	Verification Issues: Evaluation of Incident Detection Systems	74
4.6	Operational Field Tests	76
4.6.1	TRANSCOM TRANSMIT Project	76
4.6.2	I-880 Field Experiment: Incident Detection Using Cellular Phones	77
4.7	Summary	78
	Review Questions	79

5 Incident Duration and Delay Prediction 83

5.1	Incident Duration Estimation Models	84
5.2	Northern Virginia Case Study: Methodological Structure	91
5.2.1	Structure and Design of Survey Forms and Data Collection	92
5.2.2	Analysis of New Incident Data	98
5.2.3	Detailed Analysis	107
5.2.4	Summary of Detailed Data Analysis	112
5.2.5	Development of Incident Clearance Time Prediction/Decision Trees	112
5.2.6	Validation of Prediction/Decision Trees	117
5.2.7	Distribution Properties of Incident Duration Data Collected for Case Study	121
5.2.8	Comparison of Our Results With Previous Work	123
5.3	Incident Delay Prediction	125
5.3.1	Deterministic Queuing Diagram	125
5.3.2	Other Methods to Determine Incident Delays	127
5.4	Summary	128
	Review Questions	129

6	**Incident Response**	**133**
6.1	The Incident Response Problem	133
6.1.1	Tools	135
6.1.2	Research Needs for the Development of Incident Response Support Tools	136
6.2	Existing Incident Response Systems	137
6.2.1	Orange County, California: Caltrans	137
6.2.2	I-95 Corridor Coalition	140
6.2.3	Northern Virginia	144
6.2.4	Research on Incident Response	145
6.3	Formulation of a Response Plan	146
6.3.1	Incident Characterization	147
6.3.2	Service Identification	149
6.3.3	Agency Notification	150
6.3.4	Clearance Process	151
6.3.5	Computer Implementation of the Conceptual Computer-Based Response Plan	153
6.4	Case Study	154
6.4.1	Study Area and Response Statistics	154
6.4.2	Statistical Analysis of Resources	154
6.4.3	Resource Allocation	155
6.4.4	Implementation of Response Rule Base as Part of WAIMSS	160
6.5	Summary	162
	Review Questions	163
7	**Traffic Diversion for Real-Time Traffic Management During Incidents**	**165**
7.1	A Scenario	165
7.2	The Solution Approach	165
7.3	Traffic Diversion	168
7.4	Diversion System Architecture of WAIMSS	172

7.4.1	System Components	173
7.4.2	Diversion Initiation Module	174
7.4.3	Diversion Strategy Planning Module (Heuristic Network Generator)	175
7.4.4	Diversion Control/Routing Module	177
7.5	Functions and Theory of the Network Generator	177
7.6	Network Aggregation Models	178
7.7	Theoretical Modeling of the Network Generator	182
7.7.1	Elements and Types of Diversion Strategies	182
7.8	Estimation of Incident Impact Area	183
7.8.1	Representation of Incident Impact Area Knowledge	184
7.8.2	Estimation of Diversion Volume	186
7.8.3	Dynamic Link Elimination Concept	188
7.8.4	Proposed Approach for Link Elimination	189
7.8.5	Factors Influencing Link Elimination	190
7.8.6	Rule Base for Dynamic Link Elimination	194
7.8.7	Link Elimination Decision Making	195
7.8.8	Link Elimination Rule Structure	196
7.8.9	Link Elimination Decision Process	197
7.8.10	Cumulative Weight Function for Conflict Resolution	200
7.8.11	Rule Antecedents	201
7.8.12	Link Elimination Rules	201
7.9	Route Generation	201
7.10	Summary and Need for Further Research	205
7.10.1	Route Prioritization	205
7.10.2	Testing and Validation of Diversion Strategies	206
7.10.3	Multiple-Point Diversion	207
7.10.4	Network Connectivity and Existence of Multiple Diversion Routes	207

		Review Questions	207
8		**Online Traffic Control**	**211**
8.1		Introduction	211
8.2		Traffic Control Problems in ITS: Dynamic Traffic Routing/Assignment	212
8.2.1		Traditional Techniques	213
8.2.2		Ramp Metering Control	216
8.2.3		Signalized Intersection Control	218
8.2.4		Traffic Speed Control	218
8.3		Feedback Control Designs for Macroscopic Control Problems	218
8.4		Example Problem	222
8.5		Summary	226
		Review Questions	226
9		**Conclusions and Future Research**	**231**
9.1		Conclusions	231
9.1.1		Incident Input	232
9.1.2		Duration Estimation and Delay Prediction	232
9.1.3		Response Plan Development	232
9.1.4		Traffic Diversion and Control	233
9.2		Future Research	233
9.2.1		Incident Detection	233
9.2.2		Validation and Elaboration of Duration Prediction	234
9.2.3		Real-World Implementation of Duration and Delay Models	234
9.2.4		Advanced Traffic Control Algorithms	235
9.2.5		Evaluation of Existing Incident Management Programs	235

About the Authors   237

Index   239

Preface

Incident management has already become an integral part of freeway traffic operations. Most of the departments of transportation have some kind of an operational incident management plan. Traffic and transportation engineers are now more frequently asked to develop or improve incident management systems. Incident management related activities have become one of the major responsibilities of these traffic/transportation engineers.

Since the conception of Intelligent Transportation Systems (ITS) in the 80s, many transportation researchers have also worked on the development of incident management models and integrated systems for real-time operations. ITS created the required infrastructure for collecting, processing, and managing real-time traffic data that can be used to develop on-line incident management strategies. This book will provide the readers with a broad picture of the overall incident management process in the context of ITS along with a quick review of the models and systems developed by numerous researchers all over the world.

This book is a direct result of the long-term incident management research efforts at the Virginia Tech Center for Transportation Research. The initial work was performed under work order #DTFH71-DP86-VA-20 given to VDOT by FHWA. Virginia Tech Center for Transportation Research was subsequently contracted by VDOT to perform the work. In addition to this initial contract, the FHWA Intelligent Transportation Systems Research Center of Excellence (RCE) program and VDOT sponsored different parts of the research described in this book. A final comprehensive report for the original project titled "Development of a Wide-Area Incident Management Expert

System" and other reports were prepared for the sponsors of these projects, portions of which are used in this book.

One of our motivations for writing this book was to make sure that the results of this incident management research became more accessible to other researchers and practitioners who might be working on similar problems. It is hoped that the discussion of actual research projects that involved the authors of this book along with a large number of researchers at the Center will make it easier to understand the concepts presented in each chapter. Most of the chapters contain portions of the models and descriptions of the Wide-Area Incident Management Support System (WAIMSS) software. The discussions on WAIMSS are based on the papers and reports written by the two authors of this book and numerous other researchers and graduate students who co-authored these papers.

We would like to acknowledge some of the most important contributors of the incident management research at Virginia Tech Center for Transportation Research and thus to this book. Dr. A. G. Hobeika initiated the incident management research at the Virginia Tech Center for Transportation Research. Thus, authors of this book would like to acknowledge his important role at different stages of incident management research at the Center. Many graduate students and researchers worked in the area of incident management research at the Center. Among them, we would like to mention S. Subramaniam, A. Narayananan, Y. Zhang, S. Jonnalgadda, N. Dhingra, V. Khrisnaswamy, and S. Mastbrook as the main contributors to the Wide-Area WAIMSS software and its models which constitute the basis of the practical applications and research results presented in this book. Each of these researchers made invaluable contributions to the development of WAIMSS and various models used by it. At certain places in the text, excerpts, models and results from these students' papers, research memoranda, and Master's and Doctoral theses are used. Appreciation is also given to all these students and researchers who worked on several aspects of research described in this book. Information for Sections 3.4, 5.7 and 9.8 was derived from the study performed for the Federal Highway Commission by the authors and their students at the Virginia Tech Center for Transportation Research.

This book by no means claims to cover everything that has been done in the area of incident management. This would be an impossible task due to the huge amount of knowledge available in the literature. It mainly represents our understanding of the incident management process which is deeply rooted in the research conducted at the Virginia Tech Center for Transportation Research. Thus, we decided to build the book around the most important research problems we faced during the development of WAIMSS. We hope that this book provides the readers with enough information to start their own research on one of the many areas discussed in the book.

Each chapter is also accompanied by a list of references that directs readers to the important papers on the topics discussed in that chapter. We hope that this will achieve one of the important goals of this book by reducing the time needed to identify important papers and reports for conducting further research in a specific area. WAIMSS source code is available from the Virginia Tech Center for Transportation Research and interested readers can download the whole program by contacting Dr. Pushkin Kachroo at pushkin@vt.edu. Thanks are due to Bekir Bartin who helped format the book with patience and to our wives who supported us during this project.

Sections 1.1-1.7, 2.2-2.5, 5.1 to 5.3, 7.3-7.10, 9.1 and Chapters 3 and 6 are modified from the final report "Wide-Area Incident Management Expert System" by Kachroo et al. (1997) and submitted to the Federal Highwy Administration (FHWA). The study presented in this report was was performed at the Virginia Tech Center for Transportation Research under work order #DTFH71-DP86-VA-20 given to the VDOT by the FHWA.

Some of the material presented in Chapters 3, 5, and 7 is also taken from the following papers:

1. Ozbay, K., A. Narayanan, and Jonnalagadda, "Wide-Area Incident Managemnt Support System (Waims) Software" *Proc. of the Third Annual Congress on ITS*, Orlando, Florida, October 13-18, 1996. Copyright permission granted by ITS America.

2. Ozbay, K., A. G. Hobeika, S.Subramaniam, and V. Khrishnaswamy, "A Heurisitc Network Generator for Traffic Diversion During Non-Recurrent Congestion," *Transportation Research Board, 73rd Annual Meeting*, Washington, D.C., 1994. Copyright permission granted by the Transportation Research Board, National Research Council.

3. Ozbay, K., A.G. Hobeika, and Y. Zhang, "Estimation of Duration of Incidents in Northern Virginia," (reprint#971293), *Transportation Research Board 73rd Annual Meeting*, Washington, D.C., 1997. Copyright permission granted by the Transportation Research Board, National Research Council.

1

Introduction

1.1 Highway Congestion

Heavily congested urban highways in the United States have long been recognized as a severe problem, and a look into the future is not very encouraging. It appears that the daily struggle of millions of commuters and business travelers on clogged roadways, especially in metropolitan areas, will continue well into the 21st century. Congestion can be classified as recurrent and nonrecurrent. Recurrent congestion is caused by the peak hour traffic demand exceeding the available roadway capacity. Recurrent congestion is a known occurrence that can be addressed by employing different measures ranging from the building of new roads to ride-sharing programs. Nonrecurrent congestion is largely produced by traffic accidents, such as vehicle disablements and flat tires, and is the major cause of the decline in mobility in the United States. There is also a symbiotic relationship between congestion, both recurrent and nonrecurrent, and traffic accidents. Congested traffic conditions are one of the main reasons for traffic accidents; those accidents cause more congestion, which, in turn, causes more accidents. This vicious cycle is a major problem that threatens the mobility and the safety of the nation.

To make things worse, both types of congestion have ceased to be exclusively an urban problem and have become a suburban concern as well. Travel congestion on the urban interstates is increasing. The amount of peak-hour travel on urban interstates under congested conditions grew to 69% of overall peak-hour urban interstate travel in 1993, up from 52% in 1980 [1]. The drastic increase in highway travel in metropolitan areas, mainly due to the changes in urban development trends, employment, population characteristics, and automobile use, is the root cause of this ever growing congestion problem. The following changes, which are more obvious in suburban areas, are stated in many reports dealing with congestion and its effects on the nation's mobility [1–4].

- *Trends in urban development due to steady job and population growth in the metropolitan areas since the 1970s.* People living in metropolitan areas in the United States now constitute over three-fourths of the U.S. population [2]. Suburban areas took the lion's share of this population increase, which is mainly due to considerable growth in new job opportunities in those areas. In fact, "three-fourths of the overall growth in metropolitan areas has occurred in suburban areas" [2].
- *Changes in the work force.* "Two-thirds of the adult population in the United States are employed" [2], and 60% of the female population are now employed [4]. More importantly, since the 1970s, the rate of increase in employment, drivers and workers is more than three times that of population growth [4].
- *Changes in the number of vehicles and drivers.* Since 1969, the number of personal vehicles has increased 143%. Currently, 68% of households have more than one vehicle. The most dramatic growth in household vehicles occurred between 1969 and 1977, and has steadily increased since then. In 1950, 57% of the driving-age population were licensed to drive a motor vehicle. By 1993, 87.6% of the driving-age population were licensed [4].
- *Changes in the auto usage and travel patterns.* Commuting to and from work and family and personal business account for half of total personal travel. Ninety percent of all person-miles are in privately owned vehicles. Between 1983 and 1995, average work travel time in minutes increased by 13.7% [4].
- *Changes in land use.* The continued spatial expansion of metropolitan areas naturally increases auto dependency. In larger metropolitan areas, more than half the number of work trips are made between suburbs rather than between a suburb and the central business district [4].

Due to the high concentration of people and jobs in urban and suburban areas and the increasing use of automobiles for a variety of daily activities, vehicular miles of travel have steadily increased over the years. Since 1970, travel in urban areas has doubled, and travel on urban freeways has tripled [5].

It was recently reported that travel has, in fact, grown at approximately 3% per year over the past 25 years [6].

Congestion is the consequence of increased levels in demand and limited growth in capacity. The slow growth in highway capacity is mainly due to the decline in new highway construction. That decline, in turn, is due to the lack of funding, public resistance to the building new highways, and environmental concerns. It has been recently reported that growth in travel demand has outpaced the investment in highways. Capital spending on roads and bridges has dropped substantially in the last 20 years, and maintenance expenditures have been considerably reduced. The current road system, which is aging and deteriorating, has not increased since 1970. Thirty-three percent of the nation's estimated 575,988 bridges are structurally deficient or functionally obsolete [1]. As a consequence, in 1988, more than 30% of urban interstate miles were severely congested, that is, operating at levels of service D and E with volume-to-capacity ratios in excess of 0.95 during peak periods, compared to 1981, when 16% of urban interstate miles were severely congested [5].

Demographic projections are not promising either. No major study foresees any appreciable decline in the levels of congestion in the near future. Over 80% of the projected population growth is expected to occur in metropolitan areas and predominantly in the suburbs [7].

The dominance of urban travel, clearly reflected in all the government statistics, now accounts for more than 61% of the total travel in the country [4]. Congestion affects the beltways, originally built to ease congestion by allowing traffic to bypass congested cities. For example, the beltway around Washington, D.C. is one of the most congested areas in the country. As a matter of fact, in estimates of urban congestion, it is found that 18 urbanized areas exceed the threshold of 13,000 daily vehicle-miles, indicating area-wide congestion conditions on the urban freeway system [8]. The peaks in the morning and in the afternoon are increasing as a total percentage of the total daily traffic volumes; in large metropolitan areas, congestion is prevalent for four to eight hours, not just during the traditional peak hours [1].

The economic effects of congestion are also growing. Two quantities closely related to congestion are delay and wasted fuel. Those factors have great economic impacts on the users and consumers. For example, congestion on the freeways greatly affects the trucking industry, which carried 40% of all domestic freight and accounted for 78% of all domestic freight revenues, amounting to $240 billion in 1988 [2, 9]. Overall truck travel, including combination trucks, has increased over 280% since 1970. The impact is significant because the American economy is dependent on trucking, and any increase in transportation costs due to congestion is ultimately passed on to consumers.

As it is clearly stated in [2], more than any other time in our history, U.S. industries depend on the use of overseas part suppliers for the just-in-time manufacturing process. Moreover, it is clear that the nation's growth is and will be mainly in the service sector, high-technology industries, and foreign trade. Production of manufactured goods is already geographically dispersed, and the need for fast, efficient, and reliable delivery is continuously increasing [10]. New computer manufacturers located at remote corners of states such as North Dakota, Texas, and Virginia are good examples of this new economy: there will be less bulk freight, more small shipments, and increased demand for individualized freight services. That new trend requires carriers to provide faster and more efficient delivery services, and it is clear that congestion is making it costly for U.S. manufacturers to meet those service and productivity requirements.

A 1990 study by the Federal Highway Administration (FHWA) estimated that the total cost of congestion for the 50 urban areas studied was $43.1 billion. That figure represented a 10% increase ($39.2 billion) in the economic impact of the congestion since 1989 [8]. According to the same study, 13 urban areas had costs of $1 billion or more.

Other related studies have estimated that by 2005 urban congestion costs could rise as high as 8 billion vehicle-hours and $88 billion in wasted time and fuel [11, 12] (Figure 1.1). On the basis of past growth patterns and future trends, it seems quite safe to conclude that congestion will continue to grow, especially in urban areas. Congestion represents a serious and real threat to the productivity and competitiveness of the nation's economy, to the environment, and to the quality of life in both urban and suburban areas. The recognition of that potential threat by the congestion fueled the recent interest in reducing the effects of highway accidents through the use of advanced technologies and new programs. Those new efforts can be considered a sign of the importance attributed to the goal of reducing congestion by addressing the traffic incidents more efficiently.

1.2 Impact of Incidents on Highway Congestion

As previously stated, congestion is classified into two major types: recurring congestion, which is caused by the high volume of traffic on the highways, and nonrecurring, or incident congestion, which is caused by traffic accidents and other incidents.

An estimated 60% of vehicle-hours lost due to congestion is attributed to incidents. In a 1987 FHWA study, "incident congestion cost 1.3 billion vehicle-hours of delay and $10 billion in wasted time and fuel" [2]. For large

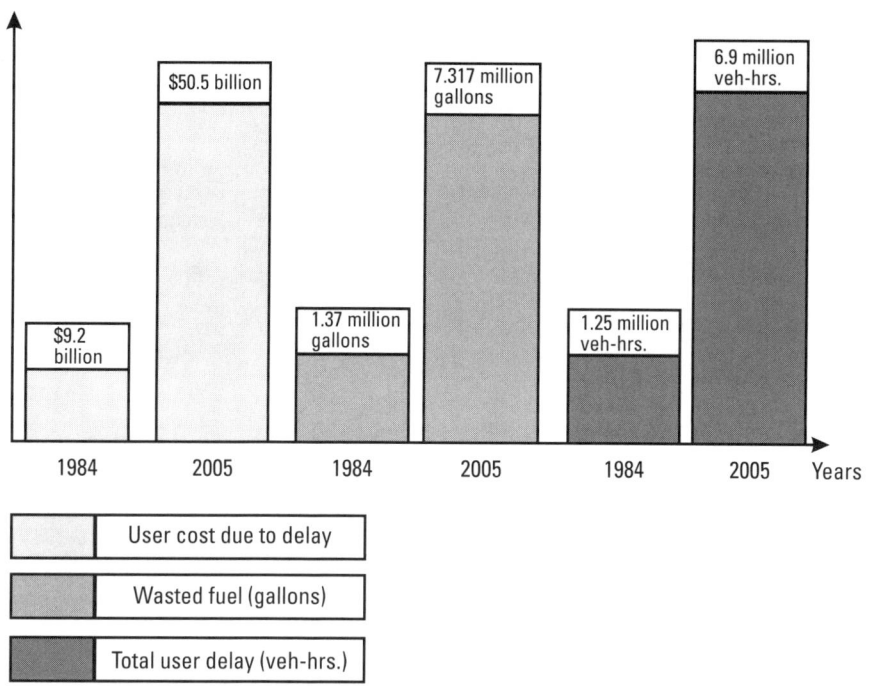

Figure 1.1 Cost attributed to congestion caused by incidents [11]. (©1989 Institute of Transportation Engineers. Reprinted with permission.)

metropolitan areas such as Los Angeles and New York, incident congestion is estimated to cost more than "$1.2 billion per year, or about $100 per person per year" [2]. More recent studies give even higher cost figures due to congestion. The total cost of congestion for the 50 urban areas studied in [8] increased by 10% in economic value between 1989 and 1990, to $43.2 billion from $39.2 billion. Representative findings of a 1990 study are shown in Table 1.1.

As Table 1.1 indicates, for every city shown the incident costs in terms of fuel and delay account for almost 60% of the overall delay. While that percentage is consistent with previous studies, the total congestion costs estimated in this study present the drastic impact of the congestion problem in terms of costs, since the overall amounts appear to be even higher than in previous studies. Moreover, the Federal Highway Administration estimates that by 2005 incident-related congestion will constitute 70% of all urban freeway congestion, a 10% increase in just 15 years [12].

Table 1.1
Component and Total Congestion Costs by Urban Area for 1990 (*Source*: Based on the data from [8])

Urban Areas	Annual Cost Due to Congestion ($ Millions)					
	Recurring Delay	Incident Delay	Recurring Fuel	Incident Fuel	Delay and Fuel Cost	Rank
Los Angeles, CA	3,000	3,530	530	620	7,680	1
New York, NY	1,950	3,630	350	640	6,570	2
San Fran- Oak, CA	1,050	1,330	190	240	2,810	3
Washington D.C.	760	1,260	130	220	2,370	4
Chicago, IL	900	1,040	160	190	2.290	5

1.3 Incident Types and Impacts

A good understanding of incidents and their impacts on the traffic flow is necessary to better appreciate the importance of incident management programs. Although the numbers presented here that describe different incident types and the duration and delay caused by each incident type are adopted mainly from studies by [2] and [13], they are consistent with other, similar studies recently conducted in several places in the United States [14, 15]. Several researchers have also studied the variables that affect the incident duration and delay for a given incident and proposed analytical models. The primary finding of those studies was that delay due to an incident is a function of the type and duration of that incident, the number of lanes blocked, the number of vehicles involved, and the involvement of a truck [16, 17].

"In general, 70% of all highway incidents are recorded by police and incident management agencies. The other 30% are minor incidents with little or no impact on traffic and as a result are not reported" [2].

The large majority of the recorded incidents, "about 80%, are categorized as vehicle disablements" [2]. Disablements are cars and trucks that have run out of fuel, have a flat tire, or simply have broken down and are abandoned by their drivers. "Eighty percent of vehicle disablements are immediately moved to the shoulder, usually by their drivers, and then cleared in about 15 to 30 minutes" [2]. Such incidents are estimated to have no significant effect on traffic flow during off-peak hours, and usually they are not even included in most of the incident duration and delay studies conducted by researchers. "During peak hours, however, they can slow down traffic on adjacent travel lanes, causing 100 to 200 vehicle-hours of delay to other vehicles" [2].

The remaining 20% of disabled vehicle incidents occur in the travel lanes and block one or more lanes of traffic. Although "most disabled vehicles are moved to the shoulder in less than 10 minutes, larger vehicles and trucks are difficult to move and end up blocking travel lanes for a longer time" [2]. According to a Cambridge Systematics study [2], on an average, it takes "15 to 30 minutes to clear the travel lane(s) of a disabled vehicle." This type of incident is reported to cause "500 to 2000 vehicle-hours of delay during peak periods" [2].

Only 10% of reported incidents are categorized as accidents, most of which are "minor collisions such as sideswipes and slow-speed rear-end collisions" [2]. According to the study conducted by Sullivan [15], around 40% of accidents occur in travel lanes, 10% on the median shoulder, and the rest on the right shoulder. In "60% of accidents, the drivers are able to move their vehicles onto the shoulder" [2]. An average accident lasts "45 to 60 minutes." During congested periods, such an accident can induce "500 to 1,000" vehicle-hours of delay [2]. The impact of accidents on traffic flow can be considerable because the presence of police cars, freeway service patrols, tow trucks, ambulances, and fire trucks reduces the freeway capacity. The freeway capacity is reduced based on the total number of freeway lanes and the number of lanes blocked due to the accident. Table 1.2 shows the overall freeway capacity available based on the total number of lanes and the number that are blocked. For example, if a shoulder accident occurs and no lanes are blocked, 19% of the freeway capacity will still be lost due to the rubbernecking of drivers passing by the incident site, and only 81% of the overall freeway capacity will be available. The important point is that, because of the rubbernecking, traffic will slow down, and a 19% reduction in capacity will occur [13]. The effect of rubbernecking is clearer for an accident on a two-lane highway with one lane blocked. In that case, instead of the 50% reduction in capacity due to the one lane blocked, a 65% reduction is observed (Table 1.2). An extra 15% reduction in capacity is clearly due to motorists rubbernecking as they pass the accident site.

Major accidents constitute only "5–15% of all accidents," and few of those are major incidents, such as hazardous material (HAZMAT) incidents, that will cause major traffic disruptions both locally and regionally [2, 14]. Catastrophic accidents can last anywhere from "12 hours to a day and require major coordination among all the parties involved, including police, fire and rescue, ambulances, and tow truck operators" [2].

According to the Cambridge Systematics study [2], "40% of accidents block one or two lanes of traffic." If they involve injuries or spills, they typically last "45 to 90 minutes, causing 1,200 to 1,500 vehicle-hours of delay"

Table 1.2
Percentage of Freeway Section Capacity Available Under Incident Conditions [13]
[With permission from the Transportation Research Board and the author.
(Original reference: Owen, J.R. And Urbanek, G.L., 1978)].

Number of Freeway Lanes in Each Direction	Shoulder Disablement	Shoulder Accident	Lanes Blocked One	Two	Three
2	.95	.81	0.35	0	N/A
3	.99	.83	0.49	0.17	0
4	.99	.85	0.58	0.25	0.13
5	.99	.87	0.65	0.40	0.20
6	.99	.89	0.71	0.50	0.25
7	.99	.91	.075	0.57	0.36
8	.99	.93	0.78	.63	0.41

According to the same study [2], an estimated "5–15% of all accidents are major accidents, each causing 2,500 to 5,000 vehicle-hours of delay." Incidents involving hazardous materials, although rare, can be catastrophic and can trigger gridlocks lasting 10 to 12 hours and causing 30,000 to 40,000 vehicle-hours of delay [2]. The remaining "10% of reported incidents are attributed to emergency maintenance work, debris on the road, pedestrians, stray animals, and other events" [2]. The impact of such incidents is minor. "Seventy percent of such incidents" occur on the shoulder, where they have minimal impact on traffic, and 30% block one or more traffic lanes for about "30 to 45 minutes and cause 1,000 to 1,500 vehicle-hours of delay" during congested conditions [2].

As previously mentioned, incidents cause bottlenecks on the roadway, slowing down or even stopping traffic. Figure 1.2 illustrates the effect of an incident on traffic using the deterministic queuing approach. When an incident occurs, it blocks one or more traffic lanes, and a queue starts building upstream of the incident due to the reduction of the capacity. The flow past the incident is a fraction of the total flow and is a function of the number of lanes available and the type of incident. Thus, quick incident clearance using effective incident strategies will reduce the clearance time and time to normal-flow conditions. Some recent studies report that losing one lane out of three causes more than a 33% reduction in capacity [15], since in addition to the physical reduction of capacity, the mere existence of the incident can further reduce the number of vehicles, i.e., capacity, that can be served. Thus, blocking one out of three lanes can reduce traffic flow by 50% and blocking two out of three lanes can reduce flow by 80% [13].The total vehicle-hours of delay experienced by the motorists in the queue are represented by the sum of the shaded and dotted areas in Figure 1.2. This delay will continue to build until the incident is cleared and traffic flow is restored to its normal conditions. If the traffic demand normally present at the incident site is reduced by diverting traffic to alternate routes, then the total vehicle-hours of delay will be minimized. This delay reduction due to the reduction of demand as a result of diversion is shown by the dotted area in Figure 1.2. The shaded area, on the other hand, is the cumulative delay if the traffic demand is reduced. However, if normal traffic demand is not reduced by diversion, additional delay shown by the dotted area in Figure 1.2 will be experienced. Following total blockage of the lanes, the capacity will be partially increased by re-opening some lanes to traffic. This capacity increase will still be less than the full capacity which will be realized when the incident is completely removed from the incident site. Following the complete clearance of an incident, the traffic waiting in queue will try to go through the incident site as soon as possible. However, the number of vehicles that can clear the incident site will be limited by the maximum capacity of the roadway at that point. This Time to Normal Flow (TNF) is also depicted as the incident recovery time in Figure 1.2. In fact, TNF needed to clear the very large queue of vehicles formed during the clearance operations is the reason for long delays following the complete clearance of major accidents on an urban freeway. In some cases, it may even take hours after the incident is cleared to dissipate this queued traffic. Once the queue is completely cleared, the traffic is considered to have returned to its normal flow conditions. Thus, quick incident clearance using effective incident management strategies will reduce the queue and the time to return to normal flow conditions.

1.4 Incident Management

Incident management is the coordination of activities undertaken by one or more agencies to restore traffic flow to normal conditions after an incident has occurred. A well-organized and coordinated incident management operation will reduce the cost of the incident in terms of delay and wasted fuel. Incident management programs that are designed to manage incidents and ease incident congestion are in place in several cities in the United States. The process of incident management consists of four sequential steps: incident detection, response, clearance, and recovery (Figure 1.3), described below [2, 18].

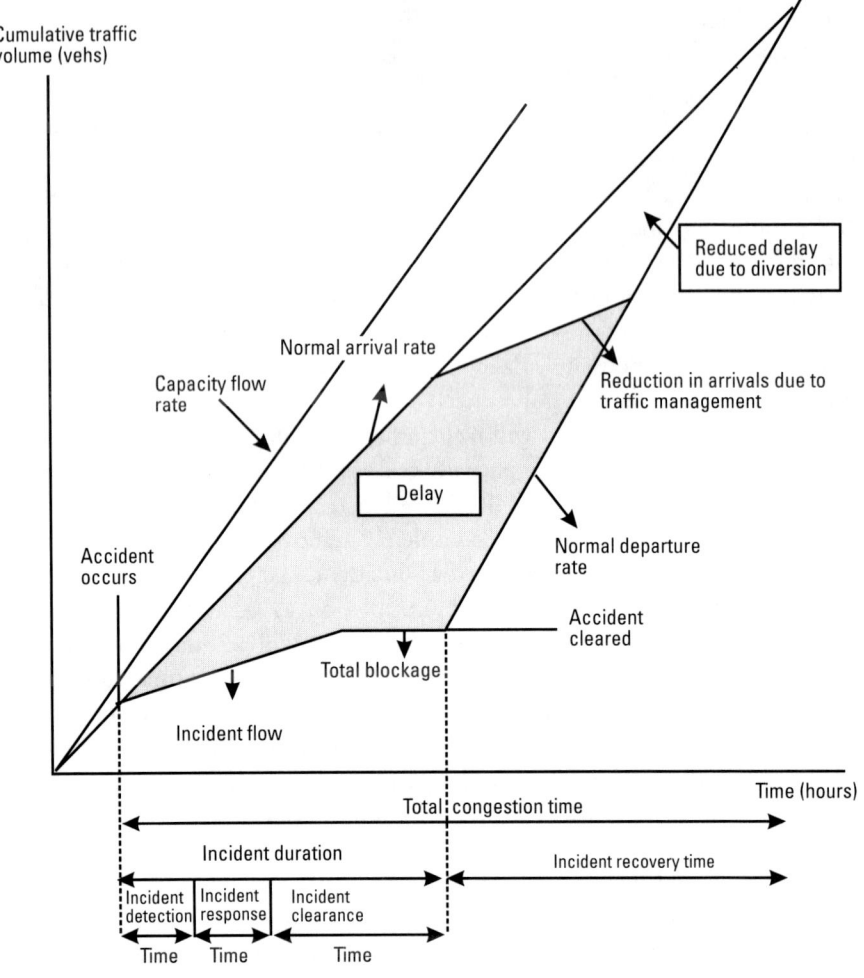

Figure 1.2 Estimation of queuing delays during an incident [based on [2] (Rover, 1990)].

- *Detection and verification.* "Detection is the determination of the occurrence of an incident. Verification is the determination of the type and location of the incident" [18]. The incident detection time of most major incidents is estimated to be between "5 and 15 minutes" [2]. The most reliable way of incident detection is routine police patrols or, where available, service patrols, as shown in recent studies [19]. An estimated "one-third to one-half of reported incidents are detected by police and freeway service patrols" [2]. In many states, including California and Virginia, freeway service patrols already exist

Introduction 11

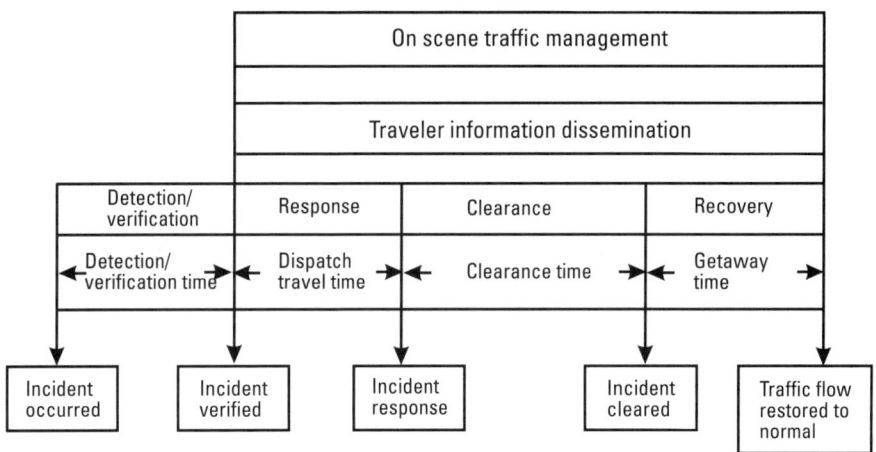

Figure 1.3 Steps of incident management [14, 20].

and perform efficiently. The rest of the incidents are reported to police from "cellular phones, two-way radios from public agency and police vehicles, and roadside call boxes" [2, 18]. In fact, in many urban areas, cellular phone calls appear to be one of the most effective ways to detect incidents. In many cities, traffic sensors such as loop detectors are installed to monitor traffic flow. The traffic flow information obtained from the detectors can also be used to automatically detect changes in flow due to incidents. Increasingly, incident detection in many highly congested urban highway corridors and tunnels is achieved through closed-circuit television (CCTV) cameras. For example, in Virginia, the traffic management center (TMC) in Arlington heavily relies on surveillance cameras installed at strategic locations of the traffic network for incident detection and verification. The effective coordination of incident information ensures timely and reliable incident detection and verification.

- *Response.* Response is the "activation, coordination, and management of appropriate personnel and equipment to clear the incident" [18]. The state police or highway patrol normally have the responsibility of confirming an incident, assessing what needs to be done, and requesting help. Typically, police dispatchers coordinate communications throughout the incident clearance operations. TMCs, however, are emerging in many cities to coordinate traffic and incident communications. Some areas and transportation corridors such as northern Virginia and the Interstate 95 (I-95) corridor from Virginia to Maine have also developed traffic diversion plans for major incidents that have regional effects.

- *Clearance.* "Clearance of an incident is the safe and timely removal of the incident and termination of the incident conditions" [18]. Around 80% of the incidents in urban areas are minor and do not need towing [18]. Usually, the police officer on the incident scene diagnoses the problem and decides what kind of towing equipment to request. The majority of freeway incidents are cleared by private tow truck operators formally or informally listed with the police departments. The police officer on the incident scene oversees the operation to make sure the incident is safely and quickly cleared. Regular training of police officers dealing with incidents ensures a successful and quick clearance of incidents.

- *Recovery.* Once an incident is completely cleared, traffic should return to normal. Time for the traffic to return to normal is considered the end of the incident. The major goals of the recovery phase are to "restore normal traffic flow conditions, to prevent more traffic congestion as new vehicles are added to the queue, and to prevent the congestion from affecting other portions of the metropolitan traffic network" [2]. Traffic management is the tool used to achieve those recovery goals. Traffic management is still in its infancy, and more research and development are needed to improve its efficiency.

1.5 Agencies Involved in Incident Management

When an incident occurs, various agencies respond. An efficient incident management process is one in which all involved agencies coordinate their operations to minimize congestion and delay caused by incidents. The various agencies involved and their roles can be described as follows [14, 20].

An incident is detected by any of the following agencies: freeway service (in cities that have them), city or state police, the department of transportation (DOT), or an aerial surveillance crew (where available). When an incident is reported either by commuters or by an agency, it has to be verified by the police or, sometimes, by a private agency such as a freeway service. Depending on the type and severity of the incident, various other agencies are called in. In modern incident management practice, the process of calling in different agencies generally is handled by a dispatcher at the traffic operations center (TOC). In the case of a disabled vehicle, the police usually call in a tow truck operator to clear the vehicle. For an accident involving injuries or fatalities, fire and rescue units and medical evacuation (MEDVAC) units respond to the incident along with tow truck operators. In the case of an incident involving a truck carrying hazardous materials, a HAZMAT management team or the Environmental Protection Agency (EPA) responds to handle the incident. If any roadway structures are affected by the incident, structural engineers are called to inspect and determine damage to the structures. During any incident, the whole operation is coordinated by DOT officials located at the TOC, who have information about the availability of resources. The DOT also handles variable message signs (VMSs), also called changeable message signs (CMSs), and the implementation of diversion routes as needed.

Figure 1.3 depicts steps of the incident management process. An important aspect of incident management is dissemination of motorist information. Motorists who receive timely and accurate incident information might decide to use alternative routes or to postpone their trips. Thus, reliable information about incident duration and severity can reduce demand by reducing congestion. However, timely dissemination of reliable incident information is crucial. Today, many major metropolitan areas in the United States have an incident management program in one form or another. Although incident management programs differ in several ways, all have adopted the overall strategy briefly described here. In addition to individual incident management programs, there are several regionwide incident/traffic management programs, such as TRANSCOM in the New York–New Jersey metropolitan region and the I-95 corridor coalition in the Northeast corridor from Maine to Virginia. Those regionwide incident management programs have been created due to the recognition of the areawide impacts of severe incidents ranging from HAZMAT accidents to long-term construction projects.

The major goals of an incident management program can be listed as follows [21, 22]:

- Reduction of the incident detection time;
- Reduction of the incident response time;
- Reduction of the incident clearance time;
- Reduction of the impact of incidents on peak-period traffic;
- Reduction of the impacts of construction, maintenance, and special events on traffic;
- Provision of accurate, timely, and useful traffic information to motorists.

To achieve those goals, the technologies and strategies listed in Table 1.3 have been proposed based on the information provided in [21, 22].

1.6 Incident Management Process

The steps involved in incident management are coordinated in a parallel, non-sequential fashion. The following are the major steps in the sequence of incident management [14, 20]:

Table 1.3
Strategies and Technologies for Incident Management Programs [21, 22]

GOALS	Examples of Strategies / Technologies
Reduce the incident detection time	Traffic Surveillance and Detection System Closed Circuit TV (CCTV) Cameras Incident Hot Lines Highway Patrol Traffic Management Centers
Reduce the incident response time	Improved Inter-Agency Communication Increased Number of Service Patrols Increased Number of Tow Truck Operators Improved Incident Response Procedures Computer Based Decision Support Systems
Reduce the incident clearance time	Improved Incident Response Procedures Increased numbers of Incident Response Teams Up-to-Date Incident Response Plans Training of Incident Response Teams
Reduce the impact of incidents on peak-period traffic	Efficient Traffic Management Strategies Efficient Diversion Plans Minimal Capacity Reduction of the Freeway Reduced Demand due to Improved Traveler Information
Reduce the impacts of constructions, maintenance, and special events on traffic	Improved Procedures Effective Public Information Effective Incident / Traffic Management Measures
Provide accurate, timely and useful traffic information to motorists	Reliable and Accurate Forecasts / Traffic Information Highway Advisory Radio Variable Message Signs (VMS) & Improved Procedures for their use for IM Improved Communication with Media (Commercial Radio and TV Stations)

1. *Devise a response plan.* Based on the type, nature, and severity of the incident and the prevailing environmental conditions, a preliminary response plan for the incident is devised (and revised when more information becomes available). The preliminary plan includes the following:
   - Agencies to respond;
   - Equipment to be provided by each agency;
   - Diversion routing plan based on historical network-wide traffic conditions and real-time data, if available;
   - Notification of back-up response needs, when needed;
   - Dissemination of incident and traveler information to the media and travelers, as appropriate.
2. *Predict incident duration.*
3. *Determine need for diversions.*
4. *Select feasible alternative routing plans.*
5. *Devise a modified response plan.* This plan suits updated incident information and alternative routing, if chosen.
6. *Recommend a complete incident response plan.* This plan includes alternative routing, messaging for VMSs, and dissemination of information to the media.
7. *Continue to update prediction of the impact of the incident.* Updating is based on additional information about the incident.

Figure 1.4 shows the sequence of the incident management process.

1.7 Problems in Incident Management

This section presents some of the major problems of incident management identified in [14]. The major problem in incident management is the effective coordination of agencies involved in the incident management process. This is not only a multiagency problem but also a multi-jurisdictional problem that requires careful planning by all the involved parties. The success of any incident management program depends on the development of effective and clear incident management guidelines and procedures that are accepted and understood by all the players. That, in turn, requires continuous communication among the players to improve and clarify existing procedures and to strengthen the web of cooperation that is vital to a seamless incident management process.

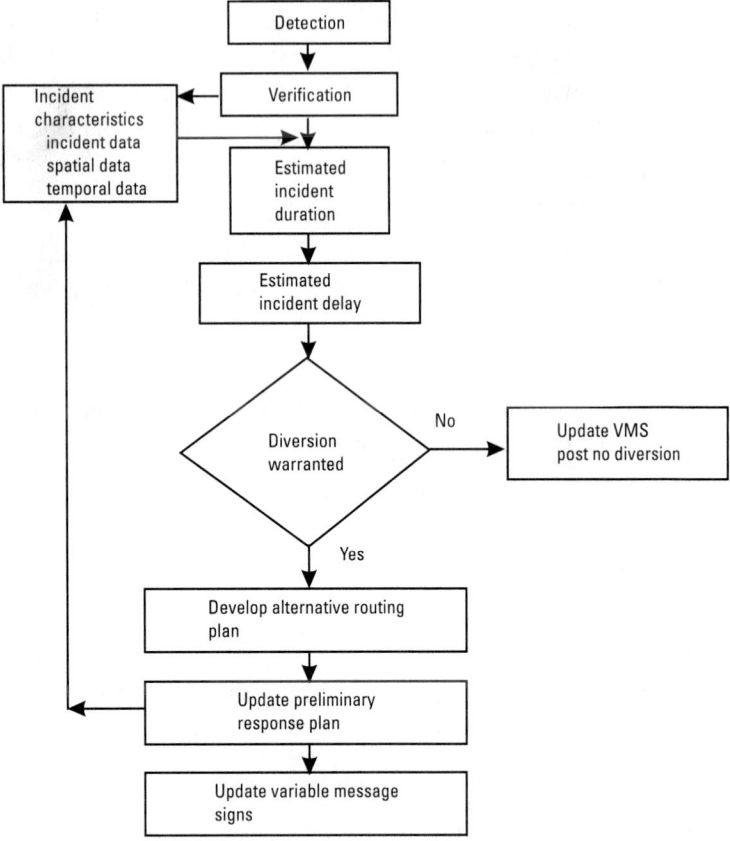

Figure 1.4 Sequence of the incident management process [14, 20].

The lack of such a mechanism that ensures interagency cooperation and communication is the main reason for the failure of any incident management program.

One of the main activities in incident management is the proper assessment of resources needed to clear the incident. An effective clearance operation requires that the police officer on the scene diagnose the problem correctly, request the right equipment, and ensure that the incident management operations are carried out efficiently. Often, that is a difficult task, especially in major incidents, "such as those involving trucks or hazardous materials (HAZMAT). Truck-related or HAZMAT incidents are much less frequent, and it is hard for police to develop adequate experience unless they receive special training in the accident clearance techniques of these types of accidents" [2]. According to a study done by the American Traveling Association [2], "clearance techniques for trucks vary widely, depending on truck size, type of load, truck configuration, and potential environmental impacts. Most police departments do not provide specialized training in incident management; as a result, tow trucks called to clear the incidents turn out to be the wrong type of equipment, and most tow truck operators do not have the expertise to handle large trucks effectively." Such problems often compound and result in the "doubling or tripling of the clearance time" [2]. HAZMAT accidents, even less frequent, present more serious problems due to the lack of experience in efficiently handling them.

When there are severe incidents in multiple sections of a major highway, diversion may also be necessary. However, systematic and up-to-date diversion plans normally are not available in many urban or suburban networks. Even when available for some subnetworks, these diversion plans are of a static nature and lack the capability to account for the real-time traffic demand.

Traffic management is considered to be the most challenging, yet it is the least developed part of incident management [18]. Most cities do not have plans for management of traffic at the incident site or for traffic diversion. Cities that have developed plans cater mainly to catastrophic events and do not have procedures for routine incidents.

An integrated system with the capability of map display, network characteristics queries, dispatching, incident duration and delay estimation, response plan generation, diversion plan recommendation and development, and interagency communication would greatly increase the efficiency of incident management and clearance [20]. Several incident management support tools have been developed to address some of those issues. In this book, we review those tools and models, with an emphasis on a computerized incident management support system for multiagency, multijurisdictional incident management support. Individual chapters discuss the results of several research efforts to the develop different aspects of a complete incident management support system. We also present some of the unique aspects of a new wide-area incident management support system (WAIMSS) we have developed to assist agencies in handling incidents.

Chapter 2 of this text is a comprehensive literature review of the general frameworks developed for the development of computerized incident management support systems. Chapter 3 describes the system architecture of a proposed incident management system. Chapters 4 through 8 describe in detail the core functional modules of an incident management system, incident detection, duration prediction and delay estimation, incident response plan development, and network generator for alternative routing, and traffic control for diversion. Finally, Chapter 9 discusses conclusions and recommendations for future work.

References

[1] U.S. Department of Transportation, "Our Nation's Highways: Selected Facts and Figures," U.S. DOT, Federal Highway Administration, Office of Information Management, 1995.

[2] Cambridge Systematics, "Incident Management," Alexandria, VA: Trucking Research Institute, Oct. 1990.

[3] Apogee Research, "Evaluating the Performance of Eight Categories of Public Works: Highways, Streets, Roads, and Bridges," in Appendix 1 of *Fragile Foundations, A Report on America's Public Works*, Washington, D.C.: National Council on Public Works Improvement, Feb. 1988.

[4] Nationwide Personal Transportation Survey (NPTS), "Our Nation's Travel," Early Results Report, 1995.

[5] Federal Highway Administration, "Highway Statistics," Washington D.C.: Federal Highway Administration, 1988.

[6] Festin, S. M., "Summary of National and Regional Travel Trends," Office of Highway Information Management (1995), 1970–1995.

[7] U.S. Department of Transportation, "Demographic Trends in National Transportation Strategic Planning Study," Washington, D.C.: U.S. DOT, Mar. 1990.

[8] Federal Highway Administration, "Estimates of Urban Roadway Congestion—1990," Washington, D.C.: Office of Traffic Management and Intelligent Vehicle Highway Systems, FHWA, Mar. 1993.

[9] Smith, F. A., "Transportation in America," Westport, CT: The Eno Foundation for Transportation, 1989.

[10] U.S. Department of Transportation, "Moving America: New Directions, New Opportunities, Statement of National Transportation Policy; Strategies for Action," Washington, D.C.: U.S. DOT, Feb. 1990.

[11] Lindley, J. A., "Urban Freeway Congestion Problems and Solutions: An Update," *ITE Journal*, Dec. 1989.

[12] Lindley, J. A., "Urban Freeway Congestion: Quantification of the problem and Effectiveness of Potential Solutions," *ITE journal*, Jan. 1987.

[13] Lindley, J. A. "A Methodology for Quantifying Urban Freeway Congestion," *Transportation Research Record 1132*, Washington, D.C.: Transportation Research Board, Jan. 1987.

[14] Kachroo, P., K. Ozbay, Y. Zhang, and W. Wei, "Development of a Wide Area Incident Management Expert System Software," (work order # DTFH71-DP86-VA26) FHWA Final Report, 1997.

[15] Sullivan, E. C., "New Model for Predicting Incidents and Incident Delays," *ASCE J. Transportation Engineering*, Vol. 123, No. 4, July/Aug. 1997, pp. 267–275.

[16] Ozbay, K. A. G. Hobeika, and Y. Zhang, "Estimation of Duration of Incidents in Northern Virginia," presented at 1997 TRB Annual Conference, Washington, D.C., 1997 (reprint #971293).

[17] Garib, A., A. E. Radwan, and H. Al-Deek, "Estimating Magnitude and Duration of Incident Delays," *ASCE J. Transportation Engineering*, Vol., 123, No. 6, Nov./Dec. 1997, pp. 459–466.

[18] Judycki, D., and J. R. Robinson, "Freeway Incident Management," Washington, D.C.: Office of Traffic Operations, Federal Highway Administration, 1988.

[19] Skabardonis, A., T. Chira-Chavala, and D. Rydzewski, *The I-880 Field Experiment: Effectiveness of Field Experiment Using Cellular Phones,* California PATH Research Report, UCB-ITS-PRR-98-1, 1998.

[20] Subramaniam, S., "Wide-Area Incident Management Expert-GIS System Development," Project Progress Report, Virginia Tech Center for Transportation Research, 1994.

[21] U.S. Department of Transportation, "Freeway Management Handbook," Report No. FHWA-SA-97-064, 1997.

[22] Roper, D. H., "Freeway Incident Management," NCHRP Synthesis of Highway Practice No. 156, 1990.

Selected Bibliography

Goolsby, M. E., "Influence of Incidents on the Quality of Service," Highway Research Record No. 349: *Traffic Flow, Capacity, and Quality of Service: 5 Reports*, Washington, D.C.: Highway Research Board, National Research Council, 1971, pp. 41–46.

Morales, J. M., "Analytical Procedures for Estimating Freeway Traffic Congestion," *Public Roads*, 50(2), 1986, pp. 50–61.

Owen J.R, and G.L. Urbanek, "Alternative Surveillance Concepts and Methods for Freeway Incident Management, Vol. 2: Planning and Tradeoff Analyses for Low-Cost Alternatives, U.S. DOT, FHWA, Report #FHWA-RD-77-59, March 1998.

Roper, D.H. "Freeway Incident Management," National Cooperative Highway Research Program, Synthesis of Highway Practice 156, Transportation Research Board, Washington D.C., 1990.

2

Review of Incident Management Systems

2.1 Introduction

Since the early 1970s, transportation departments across the United States have recognized the importance of a multidisciplinary team approach for incident management. Initially, the focus of incident management programs was on institutional issues, because of the multijurisdictional and multiagency nature of the incident management process. Once the institutional issues are adequately addressed—not always an easy task—the next step is to provide incident management personnel with efficient tools to deal with issues related to the actual management of incidents. Among those issues are:

- How to automatically detect incidents;
- Designating who should be responsible for managing different types of incidents and how to make that decision in real time;
- How to clear incidents faster;
- Whether the police officer at the incident site should have traffic engineering knowledge;
- How to coordinate the actions of different agencies in a timely manner;
- How to optimize the resource requirements for multiple incidents;
- Determining the best way of developing detours around incidents;
- Deciding what kind of information should be given to the public.

What is clear is that a centralized system is needed to address these important issues efficiently. The system should be located in a Traffic Management Center (TMC), and it should be easy to use and take advantage of an emerging advanced communication and computing infrastructure. More importantly, the system must provide a common incident management architecture to support real-time data sharing and multiagency incident response through a multiagency network. Both the FHWA and individual states recognize the need for the development of such systems to help highway departments, police agencies, and other related agencies to manage incidents efficiently. Subsequently, several research projects have been funded. As a result of those research and development efforts, computer-based support incident management systems have been developed and implemented.

This chapter reviews computer-based support systems designed to aid incident management operations. The discussion covers both existing systems and conceptual architectures proposed for such systems. We limit our discussion to complete systems and do not discuss research dealing with individual components of incident management problem. The literature related to individual submodels of the incident management process, such as duration/delay estimation, incident response, and traffic diversion are discussed in later chapters, where submodels and their roles in the context of incident management are described in detail.

As part of the section on proposed frameworks, blackboard architecture is discussed here in detail as a suitable problem-solving approach for incident management. Thus, this chapter is dedicated to the discussion of the conceptual frameworks for incident management and various systems developed for actual implementation of those frameworks.

2.2 Proposed Implementation Frameworks for Incident Management Support Systems

Most studies on incident management frameworks stress the actual sequence of operations and deal with specific components of the problem such as duration prediction, delay estimation, and traffic diversion. Issues such as the evolutionary nature of incident management operations, the presence of multiple users, and the need for cooperative decision making have often been overlooked. The emphasis of Section 2.2 is on a description of studies aimed at developing a complete framework for the complex incident management process. The demands of the system shown in Figure 2.1 can sometimes be overwhelming. The TOC exchanges data and information with different elements of the traffic system, including the surveillance, control devices, agencies, and travelers. The

Review of Incident Management Systems

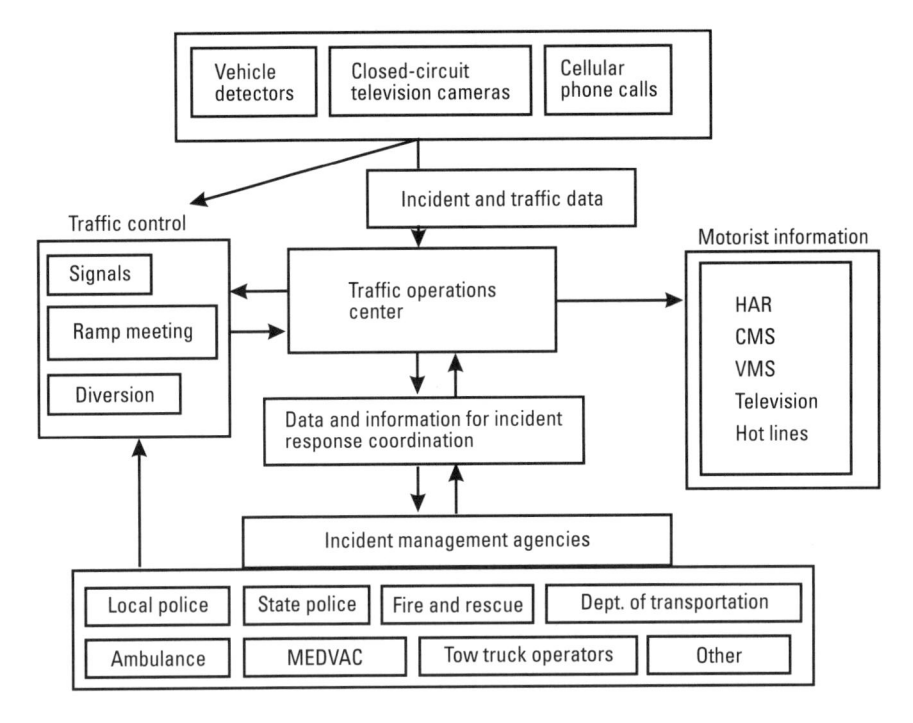

Figure 2.1 Elements of a traffic system for the incident management process.

role of a complete incident management framework is to obtain information from different sources, to process it, and to disseminate it to appropriate elements in the system. The process flow is shown in Figure 2.2. It is important to note that the process flow should be seen as a continuous feedback loop. The system developed should be able to obtain data continuously and should revise and update its decisions based on the new data. Also, coordination of various agencies' responses and traffic control strategies is an ongoing element of the overall process. In brief, the process flow should not be seen as a sequential mechanism but rather as a continuous evolutionary mechanism.

The few studies that have identified those issues and the proposed frameworks that allow a more accurate representation of response operations after an incident are discussed next.

2.2.1 System Requirements and Characteristics

Incident management operations require extensive manipulation of data obtained from traffic sensors and incident management personnel using expert knowledge. Expert systems appear to be the perfect tools to tackle the

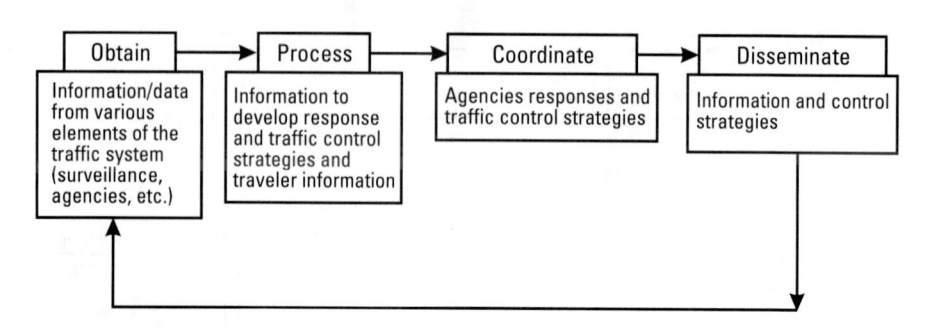

```
┌─────────┐    ┌─────────┐    ┌──────────┐    ┌────────────┐
│ Obtain  │───▶│ Process │───▶│Coordinate│───▶│ Disseminate│
├─────────┤    ├─────────┤    ├──────────┤    ├────────────┤
│Information/data│Information to│Agencies responses and│Information and control│
│from various │develop response│traffic control strategies│strategies│
│elements of the│and traffic control│          │            │
│traffic system│strategies and│          │            │
│(surveillance,│traveler information│       │            │
│agencies, etc.)│         │          │            │
└─────────┘    └─────────┘    └──────────┘    └────────────┘
```

Figure 2.2 Flow of actions during the incident management process.

data-intensive problems that require expert knowledge for developing real-time strategies. Ritchie and Prosser in [1, 2] propose a real-time knowledge-based decision support architecture for advanced traffic management. A major component of their proposed system concentrates on providing decision support for traffic management personnel for addressing nonrecurrent congestion in large or complex networks. They also identify the important characteristics and requirements for a real-time knowledge-based decision support system [1].

- *Truth maintenance.* During the incident management process, as data enter the system, the validity of any number of facts may change with time. The true state of the system needs to be monitored continuously. That is especially true for detection, verification, and characterization of an incident. For example, when a traffic disturbance is first detected through a cellular phone call from a motorist, the existence of the incident is not known for sure. The operator at the Traffic Operations Center (TOC) needs to have more information to classify the event as an incident. It is clear that the measures to respond to the incident even after it has been verified will change based on the continuous information regarding its characteristics. Changes in the incident characteristics directly affect the resources needed to clear the incident. Thus, a change in the truth value of a fact higher in the hierarchy affects the truth values of other events.

- *High performance.* Incident management is a real-time process that requires relatively short response times, usually to process rapidly changing data. Thus, the developed system should have the capability to change its recommended response strategies based on the changes in data.

- *Asynchronous events.* Unscheduled events should be capable of interrupting the process, according to their importance levels.

- *External and sensor interface.* Real-time systems need to be capable of gathering data for thousands of variables from sensors or via database interfaces and to provide continually updated information to the operator. That is one of the major issues regarding incident management systems. The sheer volume of traffic sensor data makes it impossible to use systems that do not have the capability to handle high volumes of data in real time. In the context of interfacing of data, real time is defined in terms of seconds, not minutes. An operation as simple as updating the traffic volumes on the screen might take minutes, however, if the tools chosen are not appropriate.

- *Uncertain or missing data.* The developed system should be capable of identifying and appropriately processing uncertain or missing data due to the inaccuracies or malfunction of traffic detectors. That raises the question of malfunction detection and using historical data when some of the detectors are not working properly. On the other hand, the system should be designed to warn the operator at the TOC about sensor malfunctions and the possible consequences of using historical data.

The system proposed by Ritchie and his colleagues in [1–3] employs a *blackboard architecture* to address those requirements by integrating knowledge sources at different agencies and designing a networked real-time knowledge-based expert system (KBES) running initially on separate microprocessors at each agency (Figure 2.3). To enable interactive data input to the central KBES and to permit viewing of various corridor status reports, some agencies are provided with networked terminal displays. Communications between the central KBES and the agency KBES occur via the respective database servers using the blackboard architecture. The design provides decision support to traffic management personnel through five integrated modules: incident detection, incident verification, identification of alternative responses, evaluation of alternative responses, and monitoring recovery.

Comfort proposed a similar framework for an interactive emergency information system to improve organizational decision making capacity in emergency management [4]. Although that study does not explicitly deal with incident management operations, it addresses many of the issues related to traffic management after incidents. Moreover, the architecture proposed by Comfort also employs a blackboard model to permit separate knowledge sources to interact with the global database as the time period of the emergency progresses. The design for the system concentrates on integrating the three issues in artificial intelligence: technology- knowledge acquisition, knowledge representation, and knowledge utilization:

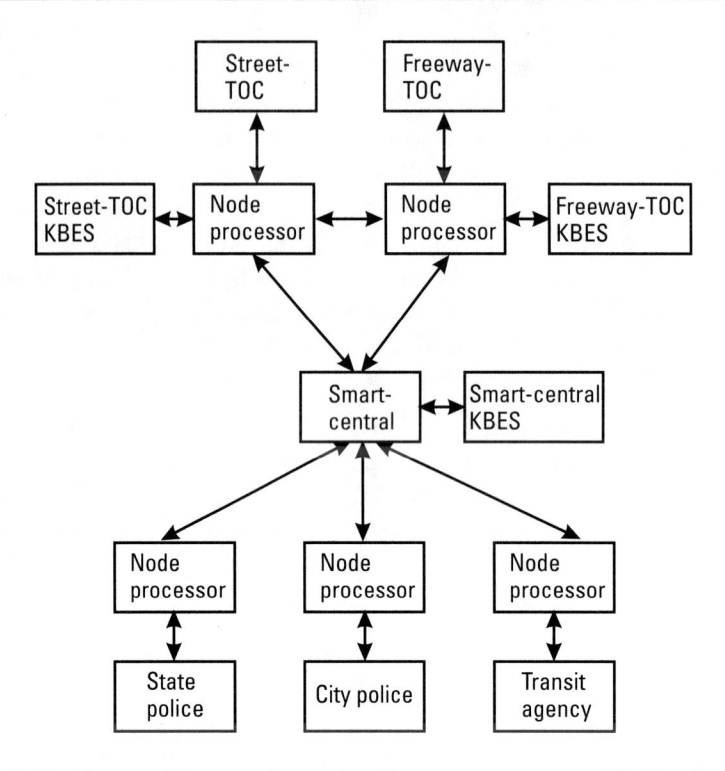

Figure 2.3 Ritchie's architecture for a decision support system [1] (Reprinted from *Transportation Research*, Vol. 24 A. NO 1, S.G Ritchie, pp. 27–37, ©1990, Pergamon Press with permission from Elsevier Science.)

- *Knowledge acquisition.* This step in design pertains to the collection and organization of information to be stored in the knowledge base. Comfort (1986) in her paper points out that, because multiple users are involved with complementary emergency responsibilities at multiple locations, it is essential that the knowledge be compiled in clear, uniform terminology and in a format understandable to all the users.
- *Knowledge representation.* Efficient representation of knowledge in the design of the emergency management system is identified as the critical issue from the perspective of the users and the design of data structures and software constructs that will be used to model the problem. Comfort (1986) provides an initial schema for the organization of the knowledge required for an emergency information system for a city jurisdiction (Figure 2.4). It assumes a blackboard architecture that permits different knowledge sources to interact with a global database during the response process. The knowledge to be stored in the

Review of Incident Management Systems 27

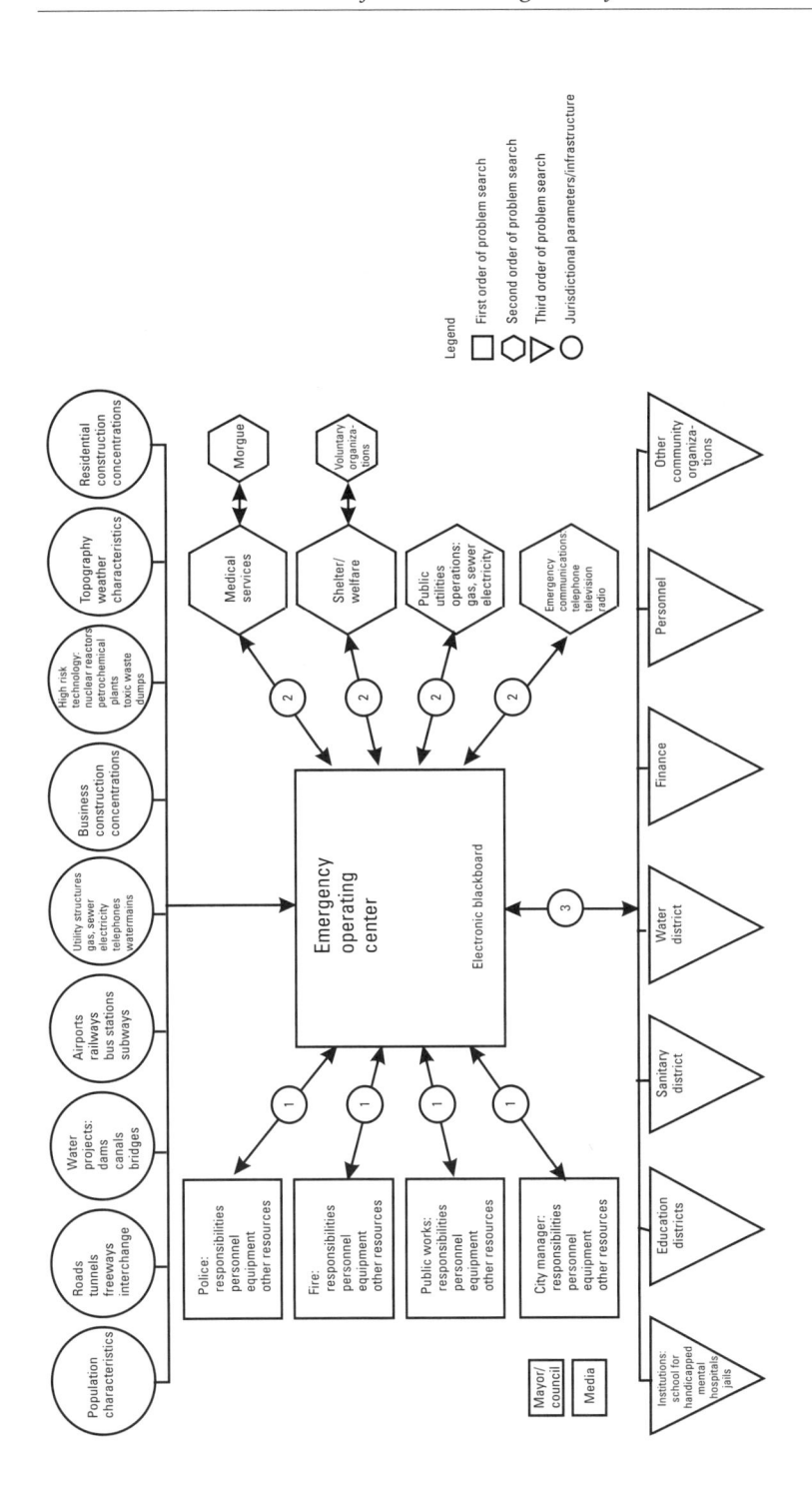

Figure 2.4 Comfort's architecture for an emergency information system [4]. (With permission from natural Hazards Center, University of Colorado, Boulder, and L.K. Comfort, *Applications in Emergency Management*, 1996.)

computer is organized into separate units corresponding to the organizations that have assigned responsibilities under the jurisdiction's emergency plan. The blackboard serves as a platform for interaction between the various organizations as they request information, enter new information, or initiate new search strategies for problem solutions.

- *Knowledge utilization.* This task involves the definition of the inference processes used to activate the information system. Comfort emphasizes the need for decentralizing the inferencing mechanisms to improve the accuracy and effectiveness of the decisions taken.

The two architectures highlight the advantages of using a blackboard architecture for organizing knowledge and providing a strategy for applying that knowledge during emergency and incident response operations. It is important to discuss the theory behind blackboard architecture for expert systems to better understand its advantages for incident management problems. The next section further reviews the blackboard architecture in terms of its potential as a problem-solving model for incident management.

2.2.2 Blackboard Architecture

The blackboard architecture was first implemented as the basis for the HEARSAY-II speech/understanding system [5, 6]. Developed between 1971 and 1976, the HEARSAY-II system responded to spoken queries about computer science abstracts in a database. Soon the blackboard architecture evolved into a robust model of problem solving, and its ability to handle complex and ill-structured problems led to the development of a number of other applications. References [5, 6] identify the following characteristics of a problem that make it an appropriate candidate for the blackboard approach:

- A large solution space;
- A variety of input data and a need to integrate diverse information;
- The need for many independent or semi-independent pieces of knowledge to cooperate in forming a solution;
- The need to use multiple reasoning methods;
- The need for multiple lines of reasoning;
- The need for an evolutionary solution.

Response operations after the occurrence of traffic incidents typically reflect most of those characteristics. The blackboard model, as its name suggests, basically tries to emulate a group of problem solvers (experts) gathered around a blackboard, solving a problem. The blackboard in this case corresponds to a shared memory that facilitates communication and cooperation among the group members. The group members are experts or sources of knowledge that contribute to the incremental development of a solution. The purpose of the blackboard is to hold computational and solution-state data needed for and produced by the knowledge sources. Essential components of the blackboard are described as follows in [5, 6] (Figure 2.5):

- *Knowledge sources.* The knowledge sources are logically independent units that together provide the knowledge needed to solve the problem.
- *Blackboard data structure.* The blackboard data structure is a global database that stores the computational and solution-state data. The knowledge sources produce changes to the blackboard that lead incrementally to a solution to the problem. The blackboard serves as a

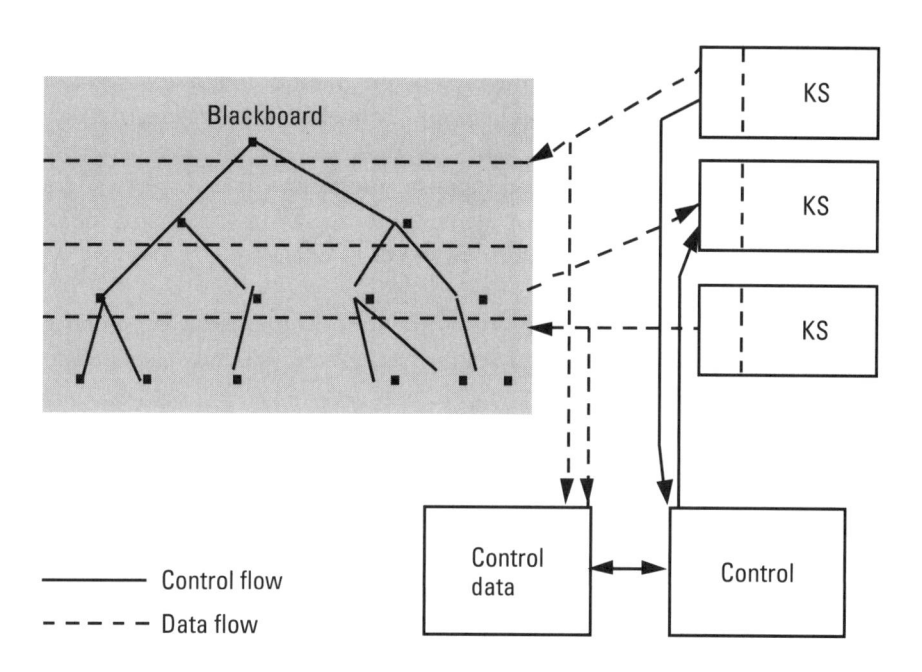

Figure 2.5 Structure of a blackboard model. (Reprinted from [5], with permission from the American Association of Artificial Intelligence ©1986.)

common source of current information about the problem and as a medium of communication and interaction between the knowledge sources.

- *Control.* The control component provides a scheduling function to determine which knowledge source to activate given the current state of the solution. The idea behind the control component is to enable the knowledge sources to respond opportunistically to changes in the blackboard architecture.

The data on the blackboard are hierarchically organized. The knowledge sources are logically independent, self-selecting modules. Only the knowledge sources are allowed to make changes to the blackboard. On the basis of the latest changes to the information on the blackboard, a control module selects and executes the next knowledge source.

It is important to note that the blackboard approach does not specify a particular method of knowledge representation and reasoning strategy. As Nii points out, the blackboard model does not specify the realization of a computational entity and is rather a conceptual entity that provides guidelines for sketching a solution to a problem [5, 6]. Hence, the design of a blackboard architecture is greatly influenced by the nature of the application problem itself. The extremely flexible nature of the blackboard architecture, in addition to its essential components, makes it an appealing model for the implementation of real-world incident management.

A number of computer-based support systems have been built to address specific aspects of the incident management process. Of those systems, the two approaches that are discussed in detail in this chapter are:

- The use of expert systems–based frameworks to provide decision support to operators through appropriate procedural and rule-based reasoning techniques;
- The use of frameworks based on geographic information systems (GISs) to provide adequate information support to operators through tools that selectively query and analyze network-related information.

Those two major approaches have been adopted for the development and implementation of incident management systems. Recently, a third hybrid approach combining those two approaches was proposed by researchers at Virginia Tech and Rutgers University [7]. This hybrid approach to the design of a computer-based incident management system is discussed in detail in

Chapter 3. Discussion of these incident management support systems presented in Sections 2.3 and 2.4 are largely based on the report by Kachroo et al. [8].

2.3 Incident Management Frameworks Based on Expert Systems

Expert systems play an important role in situations in which decision makers have to deal with cognitive overloads due to ill-structured problems, dynamic conditions, or multiple operations. In the context of traffic and incident management, an expert system can present operators with high-level analysis and recommendations concerning incident response. That reduces the operator involvement needed to focus on true operational problems. It also provides for optimal and consistent traffic management strategies, emergency response plans, and travel advisories. As a result, motorist delay and other costs associated with incidents can be reduced.

References [9] and [10] discuss the development of an expert system for the selection and diversion of routes during incident conditions. Built on Kappa-PC, the main goal of the expert system is to generate a set of alternative routes to divert traffic upstream of the incident. It first uses characteristics such as congestion levels, available capacities, and safety factors (icy bridges, height restrictions, etc.) to eliminate links and to arrive at a reduced network from which the alternative routes are generated. A user-equilibrium assignment module to predict traffic flows in the future is also incorporated into the framework. In addition, a module that calculates the clearance time for the incident based on its characteristics is included. Although the system was not tested over a real network, the study clearly demonstrated the advantages of rule-based reasoning in representing qualitative information about the network and expert knowledge in incident management. Considering the huge data requirements for a real network, the study strongly recommends the use of a workstation environment for future development. It also suggests using the Nexpert-Object expert system shell instead of Kappa-PC.

References [2] and [11] describe a real-time decision support system for freeway incident management and control. The vast amount of information flowing into the traffic control center is stated as an important reason for automation. The freeway real-time expert-system demonstration (FRED) strongly emphasizes incident detection and verification. The FRED system currently operates on a simulated freeway network in Orange County, California. For the collection of incident data, the model uses inductive loop detectors, CCTVs, CMSs, and ramp meters. FRED has the capability to handle multiple incidents and prioritize the various incidents based on their characteristics (hazardous material, accident, stalled vehicle, etc.). It also studies the delay-causing potential of the incidents to prioritize them. Having determined their importance, FRED makes the following decisions [2, 11]:

1. Determines if a major-incident traffic management team needs to be present at the incident site;
2. Provides possible CMS responses to provide information to motorists traveling on surface streets and freeways and a static alternative route to reroute traffic around the incident;
3. Suggests ramps that need to be closed based on the demands and the incident conditions.

Expert systems have also been developed to assist the teams in incident management. Gupta and colleagues [12] proposed the development of an expert system for freeway incident management on the Massachusetts Turnpike. That expert system identifies the incident type and location and the material being carried by the vehicles that are involved in suggesting the response strategies. Based on the position of the incident, the program develops displays of preplanned diversionary routes, which are checked for available capacities in real time. The KBES developed for incident management on the Massachusetts Turnpike consists of the following components [12]:

- *Incident detection and verification.* A freeway surveillance system is required to perform this task. In the system used in [12], passing motorists or police patrol are used for surveillance. Verification is once performed using information provided by a police patrol or closed circuit television cameras.
- *Classification of the incident.* Information is then provided by the police patrol on the incident characteristics, such as time, location, and the severity of the incident.
- *Notification of the incident.* This module notifies all the agencies required to clear and manage the incident site after verification of the incident.
- *Diversion module.* This module determines if diversion is necessary and checks if the preplanned diversion routes are available (i.e., if the volume-to-capacity ratio along the alternative route is less than 0.6).

This paper also explains the basic system needed for incident management. However, the different incident types considered are not complete. In addition, the routes for diversion cannot be selected anew if the preplanned routes are congested due to extraneous factors such as road construction or maintenance.

Ketselidou proposes the use of an expert system model for post-incident traffic control [13]. The model itself uses predetermined weights assigned to links based on the time of day and historical traffic volumes. Points at which diversion can be initiated and the potential destinations for particular links are also determined beforehand. A search process is then utilized to carry out an exhaustive search and come up with a set of alternative routes based on a set of criteria. The main modules in the model include [13]:

- *Data inputs.* Incident characteristics and traffic flow characteristics are employed as major inputs of the system.
- *Capacity reduction module.* This module determines the reduction in capacity of the incident link, based on the incident duration and the number of lanes blocked, and makes the decision of diversion.
- *Search algorithm.* The search algorithm selects the best route based on the preset link weights and assigns a portion of the volume to that route. The time-consuming process of selecting the best diversion route is repeated until the entire volume is diverted.

The system developed was tested on part of the test network on Long Island, New York. The test network was a freeway corridor with great potential for alternative-route assessment, since a number of freeways and associated arterials run parallel to each other. The results obtained were compared with those from a simulation package called TRAFLO (developed by the FHWA) and were found to be fairly consistent [13].

2.4 Incident Management Systems Based on Geographical Information Systems

The application of GISs to transportation problems has been drawing a considerable amount of attention over the past few years. GISs are computer-based systems that store and manipulate geographic information. Dueker and Kjerne provide a broader definition and refer to GIS as complex combinations of hardware, software, data, organizations, and institutional arrangements for collecting, storing, analyzing, and disseminating information about areas of the earth [14]. The main advantages of using a GIS for spatially referenced information are the volumes of data it can handle, the visual interface provided, and the speed with which the data can be accessed, manipulated, analyzed, and displayed.

GISs offer a number of advantages that make them suitable for transportation applications. Stokes and Marucci [15] note that GISs, by virtue of their ability to conduct spatial analyses, are ideally suited for many of the transportation management systems being conceived as part of intelligent transportation system (ITS) programs [15]. ITS programs impose new demands for collecting and integrating large amounts of information that can be referenced geographically. A GIS allows the user to integrate a variety of transportation data, such as accidents, pavement conditions, and sign inventories, and relates those data to a particular point or road segment in a spatial referencing system. Insignares and Terry discuss the advantages of applying GIS technology in traffic control systems (TCSs) and emphasize the fact that GISs provide a way to integrate the various advanced technologies being applied to traffic control [16]. Using GIS techniques in traffic control systems provides an integrating force that acts to streamline the presentation of TCS information and is a more effective means for TCS personnel to:

- Collect and categorize information provided by the TCS;
- Interpret and analyze the information;
- More quickly formulate solutions to problems that arise;
- Integrate geographically based TCS data from other agencies for the purpose of interagency analysis.

Insignares and Terry [16] highlight the importance of geographical user interfaces that can be provided through the use of a GIS and how data accumulated by monitoring devices can be assimilated geographically for later spatial correlation or merger with other geographic data for analysis. They also point out that future GIS in transportation will tend to move away from the role of analyzing historical data only and will play a greater role in the analysis of real-time and historical data to serve a more predictive role.

Clearly, the ability of a GIS to manage spatially referenced information offers an advantage for transportation response operations where current information on network conditions is vital for decision making. Siegfried and Vaidya [17] exploited many of those advantages by employing GIS technology to improve incident management operations in a Houston project. The study evaluated the use of a GIS to relate incident locations with the transportation network and to make decisions and calculations for incident management. The system developed provides computerized mapping and database management.

By using a GIS development platform, it was possible to develop interrelated maps, databases, and incident management applications. The platform also provided the ability to integrate other software for data sharing and execution through a common user-friendly interface.

The prototype applications developed for the automated incident management plan are classified into two groups. The first group pertains to incident management operations, the second to planning and analysis for incident management.

The user-friendly interfaces developed by Siegfried and Vaidya [17] for the applications can access historical or real-time traffic data to provide input for management decisions. In addition, the applications developed were integrated with PASSER-II computer software, which optimizes traffic signal roadways or an entire network of roadways for obtaining intersection operational data updates, and the emergency management software package, CAMEO, for independent execution, to perform areawide chemical emergency management. The applications for incident management operations include the following [17]:

- *Alternative routing.* Depending on the need, alternative routing can be performed over (a) a local area, using frontage roads for incident-affected freeways; (b) a wide area, using corridor arterials; and (c) a hazard area.

- *Incident response.* Incident response plans can be provided for (a) blocking of individual freeway lanes for the purpose of alternative routing; (b) automated blocking of a hazard area; and (c) automated tracking of incidents and dispatching one or more available vehicles from the incident response fleet.

- *Resource management.* Resources required for incident management operations, including police and fire stations, hospitals, and critical flood pump locations, can be managed by (a) providing spatial queries to obtain names, addresses, and phone numbers of organizations and individuals to contact when required and (b) providing spatial queries on the features, capacities, and availability of the sources.

Applications for planning and analysis for incident management were also provided to analyze roads and incidents, separately or in conjunction, and to devise plans for improving traffic flow. These applications include the following [17]:

- *Road database query.* The road database query system is developed to permit spatial queries on the roadway inventory database as well as graphical results of queries on the database. Queries can be performed on a variety of attributes stored in the roadway inventory.

- *Incident database query.* The incident database query system is designed to provide the ability to perform spatial queries on the number and types of incidents on any section of freeway. Response to some queries could be shown graphically.

The prototype applications of the system described were developed by Siegfried and Vaidya using PC-Arc/Info GIS package on a 486 IMB personal computer [17]. Arc/Info modules were used for the development and operation of those applications. Data conversion and overlay modules were required for development. The development process identifies available map databases and other issues that must be addressed prior to implementing a large-scale automated incident management plan. Most of the functions were automated using PC-Arc/Info Simple Macro Language (SML). The SML programs operate in the route module and call other modules whenever needed.

An important finding of this study was that a PC platform may not be suitable for real-time traffic management systems, such as automatic vehicle identification based on traffic monitoring, where new data is received at short intervals of a few seconds. In addition, PC-Arc/Info is slow compared with other commercially available relational databases, and SML is limited in scope and function. The study strongly recommends the use of a workstation environment and a programming language that parallels a higher level programming language.

2.5 Summary

The frameworks proposed by [1–4, 9, 11] are significant contributions because they recognize the need to support both strategy development and the various interactions at the individual and agency levels that take place during incident management. Issues related to multiple users and the need for coordinated decision making have often been neglected and still plague many incident management operations. In that aspect, the blackboard architecture provides a suitable framework for a more accurate representation of the problem. It furnishes a platform for information sharing and exchange between different working groups and allows for an incremental and cooperative development of a response strategy with input from many independent or semi-independent (agencies, telephone calls, etc.) sources of knowledge.

An important drawback of the frameworks presented here is the fact that they do not adequately address the spatial nature of the incident management process. The process in itself involves collecting and integrating large amounts of geographically referenced information (network information, agency locations, resources, etc.), but the frameworks are limited by their failure to design for the spatial component of the problem. Existing expert systems in incident management also overlook that aspect. In addition, most of these systems deal with one specific component of incident management and do not address the need for coordinated decision making. FRED, for example, focuses on incident detection and verification and does not provide adequate support for response plan generation diversion planning [11].

The system developed by Siegfried and Vaidya [17] recognizes the spatial aspect of incident management by employing a GIS. With a GIS as the development platform, the prototype application developed aids decision making by providing facilities for querying and manipulating large volumes of geographically referenced data. However, the application lacks tools for high-level analysis of the data using procedural and rule-based reasoning. The overall framework does not address the need for coordinated decision making and information sharing among the agencies involved in the incident management process.

In summary, the systems discussed in this chapter address some specific aspects of the incident management process, but they do not comprehensively support all components of the problem. Given the fact that the entire process is a cooperative effort, there is a need for support systems that address both the substance of the task, problem, and decision and the various individual and agency-level interactions that take place during the response process. The system proposed as a part of the prototype incident management support system developed at the Virginia Tech Center for Transportation Research attempts to overcome those drawbacks through the design of a hybrid group decision support system that employs both an expert system and a GIS [7, 8]. The approach and the methodology adopted in the development of such a hybrid system are described in the chapters that follow.

Review Questions

1. Identify data needs for effective management of incidents and determine the uncertainties due to the incompleteness of data.

2. Identify the agencies involved in the incident management process in your state or county and develop a process flow diagram that shows the incident management process in the selected area.

3. Identify the institutional issues involved in managing incidents.

4. Determine the pros and cons of using expert systems versus GIS for the development of a computerized incident management system.

References

[1] Ritchie, S. G., "A Knowledge-Based Decision Support Architecture for Advanced Traffic Management," *Transportation Research Record*, Vol. 24A, No. 1, 1990.

[2] Ritchie, S. G., and N. A. Prosser, "Real-Time Expert System Approach to Freeway Incident Management," *Transportation Research Record*, No. 1320, 1991.

[3] Stack, R., and S. G. Ritchie, *CALTRANS District 12 Real-Time Expert System for Freeway Incident Management*, Final Report, Institute of Transportation Studies, Irvine, CA: University of California, 1993.

[4] Comfort, L. K., "Improving Organizational Decision Making in Emergency Management: A Design for an Interactive Emergency Management System," in M. A. Marston, ed., *Terminal Disasters: Computer Applications in Emergency Management*, Natural Hazards Center, University of Colorado, Boulder, 1986.

[5] Nii, H. P. "Blackboard Systems (I): The Blackboard Model of Problem Solving and the Evolution of Blackboard Architectures," *AI Magazine*, Summer 1986.

[6] Nii, H. P. "Blackboard Systems (II): Blackboard Application Systems, Blackboard Systems from the Knowledge Engineering Perspective," *AI Magazine*, Aug. 1986.

[7] Ozbay, K., A. Narayanan, and S. Jonnalagadda, "Wide-Area Incident Management Support System (WAIMSS)," *Proc. Third Annual ITS World Conf.*, Orlando, Florida, 1996.

[8] Kachroo, P., K. Ozbay, Y. Zhang, and W. Wei, "Development of a Wide Area Incident Management Expert System" (work order #DTFH71-DP86-VA-20) FHWA Final Report, 1997.

[9] Krishnaswamy V., *Heuristic Network Generator—An Expert System Approach for Selection of Alternative Routes During Incident Conditions*, Master's Thesis, Blacksburg, VA: Virginia Polytechnic Institute and State University, 1994.

[10] Ozbay, K., A. G. Hobeika, S. Subramaniam, and V. Krishnaswamy, "A Heuristic Network Generator for Traffic Diversion During Non-Recurrent Congestion," *TRB Conf.*, Washington, D.C., Preprint No. 94, 1994.

[11] Zhang, H., and S. G. Ritchie, "Real-Time Decision Support System for Freeway Management and Control," *ASCE J. Computing in Civil Engineering*, Vol. 8, No. 1, 1994, pp. 35–51.

[12] Gupta, A., V. Maslanka, and G. Spring, "Development of a Prototype KBES in the Management of Congestion," Preprint No. 920302, Washington D.C.: Transportation Research Board, 1992.

[13] Ketselidou, Z., *Potential Use of Knowledge-Based Expert Systems for Post-Incident Traffic Control*, Ph.D. Dissertation, Amherst: University of Massachusetts, 1989.

[14] Dueker, K., and D. Kjerne, "Multipurpose Cadaster: Terms and Definitions," *Annual Convention ACSM-ASPRS Proc.*, Vol. 5, 1985.

[15] Stokes, R., and G. Marucci, "GIS for Transportation: Current Practices, Problems, and Prospects," *ITE J.*, March 1995.

[16] Insignares, M., and D. Terry, *Geographic Information Systems in Traffic Control*, ITE Compendium of Technical Papers, 1991.

[17] Siegfried, R. H., and N. Vaidya, *Automated Incident Management Plan Using Geographic Information Systems Technology for Traffic Management Centers*, Texas Transportation Institute, Research Report 1928-1F, 1993.

3

Wide-Area Incident Management Support System Software

Chapter 2 discussed the needs for a multi-user wide-area incident management system and described the capabilities of existing individual incident management systems. This chapter presents the conceptual design for an incident management decision support system that addresses the issues raised in Chapter 2. Individual components for this conceptual framework are discussed in later chapters. It is, however, extremely important to first develop a complete conceptual framework along with implementation requirements for the successful development of a complex system similar to this one. Thus, the conceptual design for the proposed system is examined here, using the necessary implementation details and eliminating the drawbacks identified in the literature review presented in the previous chapter. The discussion in this chapter covers important issues to be considered during the conceptual framework design process and proposes an integrated framework that has both an expert system and a GIS to support the wide-area incident management process. The general framework described in this chapter, called a Wide-Area Incident Management Support System (WAIMSS), is also presented in detail in [1] and [2].

3.1 Design Considerations

The main focus of this conceptual framework for the group decision support system is to accurately represent the nature of the incident management process and to address the issues requiring decision support during the incident management operations. In addition, the real-time nature of the required decision support means that the system must satisfy demands that do not exist in conventional problem domains, in which inputs are static and timely critical responses generally are not required. To satisfy the requirements of this complex problem, it is necessary to consider the following issues.

The system should:

- Provide a database management system to support the storage and retrieval of static and dynamic information pertaining to the road network and the incidents;
- Have the ability to incorporate the knowledge of experts in the field and apply that knowledge for real-time decision making with high-level analysis and recommendations;
- Contain utilities to selectively query, analyze, and manipulate information in response to the situation;
- Possess a high level of visualization capabilities that can be used to present a large amount of information in a cohesive and context-related manner;
- Be able to coordinate the efforts of the multiple agencies involved in each operation;
- Provide a platform for communication and information exchange and sharing between different working groups.

In sum, the design should serve as a vehicle for organizing the relevant knowledge for the decision-making process in a format that is readily accessed by multiple decision makers at multiple locations. Accomplishing that task will facilitate a secondary function essential to effective incident management: the cooperation and coordination of actions within and among multiple organizations with complementary incident management responsibilities.

3.1.1 Overall Concept

The overall objective of the proposed conceptual framework described in [1] and [2] is to assist various incident management personnel involved in determining the appropriate strategies to manage incidents and support execution of steps required for their implementation. Since such strategies usually call for a cooperative effort by several agencies, the framework presented in this chapter not only concentrates on addressing strategy development but also supports the various individual and agency-level interactions that take place.

Hence, it incorporates the two important facets of incident management operations [1–3]:

- *Problem content support.* Content support for a given problem domain can be described as the capability of an overall computer-based system (hardware, software application programs, algorithms, heuristics, etc.) to provide support to its users (individuals or a group) in addressing issues related to incident management, including incident duration estimation, delay prediction, clearance strategies, and diversion strategies.
- *Group process support.* The group process support is the task of the framework that facilitates and improves the dynamic group decision-making process by enhancing participation and information exchange among the different groups working to manage each incident case.

A support system designed to serve both those functions is called a group decision support system (GDSS), or simply a group support system (GSS) [3]. A GDSS is described as an interactive computer-based system that facilitates the solution of unstructured problems by a set of decision makers working together as a group. The components of a GDSS include hardware, software, people, and procedures. An important distinction from conventional support tools is that a GDSS models the people involved as an integral part of the design. Thus, a GDSS incorporates the interactions that take place among people during the problem-solving process. That aspect of a GDSS design makes it a perfect design tool to tackle a highly complex incident management problem that involves not only software and a set of procedures but also different groups of people.

To provide the required decision support and information management capabilities, the developed architecture combines the powerful spatial data-handling capabilities of a GIS with the rule-based logic of an expert system. The applications support operations at each of the participating incident management agencies. The blackboard framework, described in Chapter 2, integrates the individual applications at various agencies and provides the required group decision process support. The following sections discuss the general architecture in greater detail in terms of its individual components, namely, framework for integration, software development process, and system modules.

3.1.2 Framework for Integration

The WAIMSS employs a blackboard architecture to support incident management operations by providing facilities for spatial and temporal data analysis and a mechanism for interaction between different responding agencies. Figure 3.1 shows the overall concept behind the blackboard architecture. For simplicity, only four agencies are shown: the TMC, the DOT, the state police, and a local agency. However, the idea behind the architecture can be used to integrate all the agencies involved in the incident management process through the blackboard.

The blackboard corresponds to shared memory that is assumed to be located at the main server in the TMC. Each agency is equipped with applications that facilitate communication with the blackboard. Information that needs to be shared during the incident management process is recorded on the blackboard. The blackboard also serves as the global database for the system and helps in recording, searching, and identifying possible strategies for response to an incident. It also monitors the status of the actions taken and records the tasks that have been addressed or solved. Further, it helps in sending messages among agencies and updating displays when required.

The blackboard is organized into five distinct areas (see Figure 3.1):

Blackboard database

1. Response plan
2. Prioritized action list
3. Static and dynamic network information
4. Resource information
5. Process Logs
   • event list
   • process-ID list

Control mechanism

Monitor blackboard events
Messaging
Display updates

Traffic management center

Comprehensive application and network database

DOT

Agency-specific application and database

Police

Agency-specific application and database

Local agencies

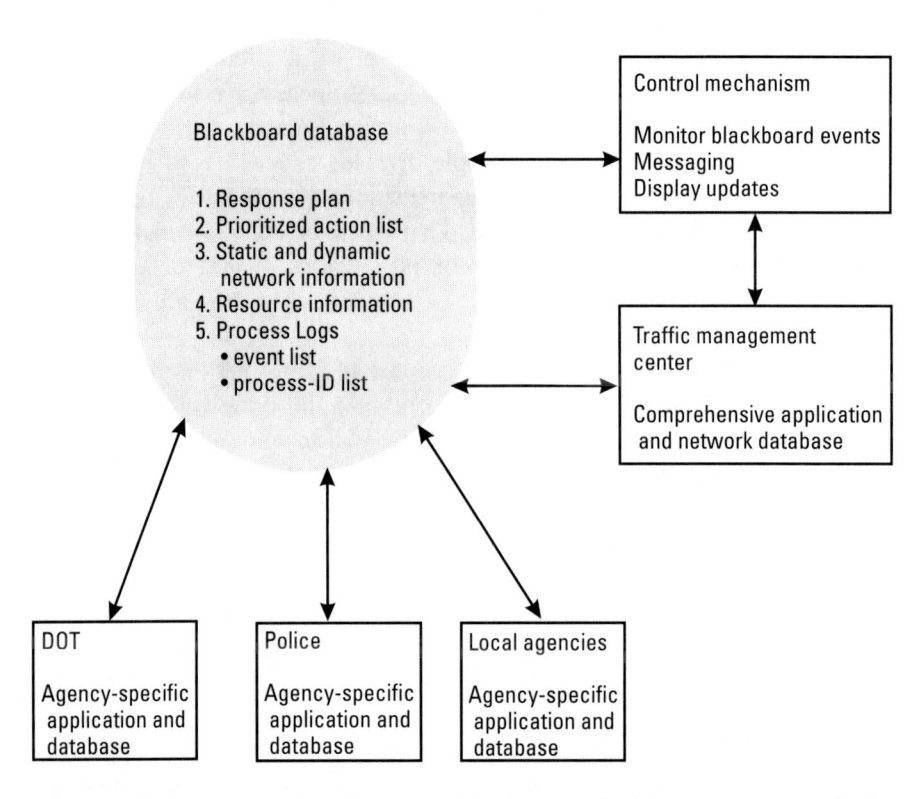

Agency-specific application and database

Figure 3.1 Conceptual blackboard framework for the group decision support system [1, 2].

- The first component is the response plan, which represents the current set of strategies chosen to manage the incident. The decisions taken for the allocation of resources, personnel, and equipment are recorded with provisions for updating the status of events as new information is reported. Once all the posted strategies are agreed on, the plan takes final shape and is considered ready for implementation.
- The second component of the blackboard fixes the agenda for actions and reports tasks in order of priority.
- The third and fourth components of the blackboard store relevant information about the traffic network and the available resources. The information may be static (e.g., network geometry, lane capacities, and link lengths) or dynamic (e.g., current link volumes, weather conditions, and available resources such as wreckers and ambulances).
- Finally, the fifth component of the blackboard framework maintains a continuous log of events after a particular incident. The log also includes a list of process IDs of the different agency applications connected to the blackboard. The IDs are used by individual agencies to pass messages.

3.2 Application Design

The proposed architecture assumes that the agencies involved in the incident management process are equipped with site-specific support tools and databases. The support tools at each individual agency provide decision support for functions addressed by that agency. The local databases at each agency store information regarding its personnel, equipment, and other resources. The TMC, on the other hand, is equipped with a comprehensive support system and a road network database, because most of the decisions regarding the response actions are taken at the TMC. In addition, the TMC application acts as a control mechanism and monitors changes on the blackboard and decides on the action agenda. To serve those two functions, each application is built as a combination of an expert system and a GIS.

The expert system uses rule-based reasoning to provide decision support for operations like duration estimation and alternative routing, while the GIS provides data management capabilities and provides an interface for spatial analysis and manipulation of network information. In addition, procedural programming capabilities are added by bridging the system with C functions. The following sections discuss in detail the software implementation and issues related to the integration of the expert system with a GIS.

3.2.1 Decision Support Modules

Decision support for traffic management operations after an incident occurs is provided through six modules, as shown in Figure 3.2. The modules are incident detection and verification, preliminary response plan generation, duration estimation, delay calculation, response plan generation, and recovery. An important point that needs to be emphasized here is the fact that the modules are not necessarily sequential. For example, duration estimation can be performed just after verification of an incident, but it also can be reestimated at a later stage during the actual incident response or recovery when more reliable and detailed information becomes available. However, Figure 3.2 is an attempt to provide a logical flow of events for the management of an incident.

The modules shown in Figure 3.2 incorporate many features that support incident response operations. The incident detection and verification module

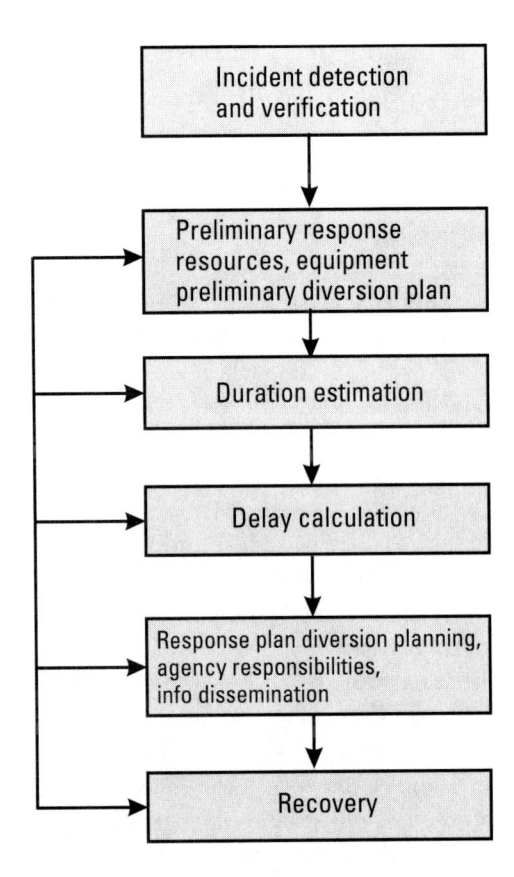

Figure 3.2 Decision support modules for WAIMSS [1, 2].

accepts operator input about the incident characteristics and requests verification from the TMC based on incoming motorist calls from cellular telephones or verification by on-site personnel. It is planned to develop the module to complement existing and ongoing research on incident detection algorithms.

Establishing a preliminary response plan is the first step of WAIMSS. The preliminary response module serves to advise clients of the system about the resources needed to clear the incident and judge if diversion would be necessary. The model is being designed to operate even when only incomplete information on incident characteristics is available. The module uses a rule base developed with Nexpert-Object expert system shell to determine the response. The guidelines for response as suggested by the module are sent to the corresponding agencies listed in Table 3.1. Once the responding agencies finalize their portion of the response, it can be updated on the blackboard and the system can proceed to predict the duration of the incident, delay, and diversion strategy with greater confidence.

3.2.2 Duration Estimation Module

The duration estimation and delay calculation modules use a rule network developed with the use of NEXPERT and C algorithms to calculate the duration of the verified incident and the resulting delay. The duration estimation module is probably the most important component that will add value to a preliminary response plan. Duration refers to the time interval that the incident is expected to last until completion of all clearance activities. The data for the duration estimation module come in from the agencies that inform the module about resource availability and resources dispatched to clear the incident. The data used by the model are shown in Table 3.2. The WAIMSS architecture allows the various agencies to update the central data needed by the duration

Table 3.1
Preliminary Response Module [1, 2]

Suggestion/Response	Agency Advised
Incident Alert	All Agencies
No. of Police Vehicles to Use	Local Police/ State Police
No. of Wreckers	Towing Company/ Local Police/DOT
No. of Ambulances	Hospitals/Rescue Squads/Red Cross/Other Medical Agencies
Diversion Strategy	Local Police/DOT/ State Police

Table 3.2
Input Data for Duration Estimation [1, 2]

Input Data	Source Agency
Incident Type, Location and Time	Any client of system
Lane Closure Information	Any client of system
Weather	Any client of system
Wreckers Used	Towing Company/Local Police/DOT/State Police
Police Vehicles Used	Local Police
Ambulances Used	Hospitals/Red Cross/Rescue Squad/Other Medical Agencies
Fire Trucks Used	Fire
Roadway Information	Static Data Base at TMC

estimation module based on the roles defined. The data are then directly accessed by the module and used for prediction. Note that estimating will continually improve as more data become available.

3.2.3 Delay Calculation Module

The predicted duration is used to determine the need to divert. It also is used by the Delay Calculation module to estimate resulting delays. In a routine operation, both those options conceivably could be applied. For example, in the northern Virginia region, an alternative routing plan is established only if an incident is expected to last over two hours, especially during or just before peak travel periods [4]. For applications in which there is a set traffic management strategy, it may be sufficient to rely on the predicted incident duration. In most cases, however, such an approximation may not lead to efficient strategies because the given duration of an incident has drastically different consequences depending on, among other factors, the time of day and the location of occurrence. An incident on a freeway lasting 45 minutes during the morning peak traffic period could cause a much longer delay than one on a primary road lasting over two hours at 10 o'clock at night. Hence, it is important to calculate the delays caused by the incident to measure the total impact on the traffic conditions over the network.

Based on the estimated duration of the incident, the delay is calculated using the deterministic queuing approach programmed in C. Network information such as the link flows and capacities are read directly from attribute tables in the Info database of the GIS package.

3.2.4 Response Module

The response module is used to finalize the response plan based on the delay, duration, and current information on resource availability. The expert system rules classify the severity of the incident occurrence based on the input parameters and indicate the equipment required to clear the incident. An important requirement for an efficient response system is the introduction of rules for checking the dynamic resource availability at the various dispatch locations around the incident and selection of a set of feasible dispatch locations. That allows the response module to select the best agency dispatch locations to respond and also suggests the quickest dispatch routes. The routing is based on minimal response time for each dispatch involved in the total response. Priority is given to response from fire, rescue, and police in light of the hazard to human life involved. This module also contains rules for media notification of the incident when required, including highway advisory radio (HAR), TV and radio broadcasts, and CMS control locations. When a selected response plan is executed, this module automatically resets itself for the next response request.

The diversion planning component of the response module uses the network generator [5] to determine if diversion is necessary and develops appropriate diversion plans. To reduce the amount of processing, an impact area is defined around the incident, and only links lying inside the impact area are evaluated for diversion planning. This module follows a three-step procedure to determine the diversion route. First, a set of heuristics examines links in the impact area to check if they are feasible for diversion. The second step uses heuristics to examine each individual link with respect to a wide array of factors, including incident characteristics, volume-capacity ratios, weather, link type, and special events to determine whether it is feasible for diversion.

The third step employs a route-generation module to find a diversion route using the feasible links. This module can use a simple shortest path routine or Kth shortest path routines to generate alternative routes using the feasible links obtained by the network generator. In addition to the real-time route generation, WAIMSS also allows the operator to query and evaluate any predefined alternative routes in a historic route database. Those routes, if available for a particular incident link, can be queried by name or related incident link. Their applicability to the current incident scenario is an operator decision made during the response effort.

In addition to its querying and display capabilities, WAIMSS also provides an interactive interface to input or update predefined routes in the database. The routes are stored as a route system in Arc/Info. A route system is defined as a collection of routes representing a common linear entity, for example, a collection of diversion routes can be called as a "DIVERSION" route system. The ability to interactively modify and update existing alternative routing plans adds a planning component to the system along with the real-time decision support capabilities.

3.3 Software Implementation

The software implementation of a complex framework such as WAIMSS presents a challenge. The most important source of implementation problems is the need for integrating different software packages that are best suited to provide the functionality required for the specific module. That problem can be circumvented by using a low-level programming language such as C/C+ but, that would require the development of several specialized tools such as GIS and the expert system shell from scratch. That would not be the most practical and economical approach for developing a reliable real-time system for incident management. Instead, specialized software packages should be chosen to implement each module described in this chapter. That guarantees the use of the best software tool available for the implementation of an individual module, because the individual software packages were developed by highly skilled software companies that spent millions of dollars to optimize and test them.

During the course of the WAIMSS research project several software packages for GIS and Expert System Shell have been considered. For GIS-related applications, Arc/Info appeared to be the most versatile and widely used GIS package for the incident management type of applications [6]. As a matter of fact, many DOTs in the United States, including the Virginia DOT, are already using Arc/Info for other GIS projects [6]. That is a considerable advantage, because they already own Arc/Info and are already familiar with it. For expert system development, we chose the Nexpert-Object expert system shell, which provides a complete package tailored for large-scale expert development projects. For graphical user interface (GUI) development, we selected Open Interface Elements toolkit, which, because of its close relationship to Nexpert-Object, is a natural tool for our software implementation efforts. The software components of WAIMSS and the way they interact with each other are shown in Table 3.3 and Figure 3.3.

Certainly, other software packages can be used to perform the same functions. However, during this research project the software packages mentioned here have proved to be the perfect tools for the most effective solution of this large, complex problem. In addition, C programming language is used to

Table 3.3
Software Components and Tools Chosen for WAIMSS [1, 2]

Functionality	Tool
Data Base Management	Arc/Info
Knowledge Processing/Inferencing	Nexpert-Object
Graphical and Procedural Programming	C
User Interface	Open Interface Elements, Arc/Info

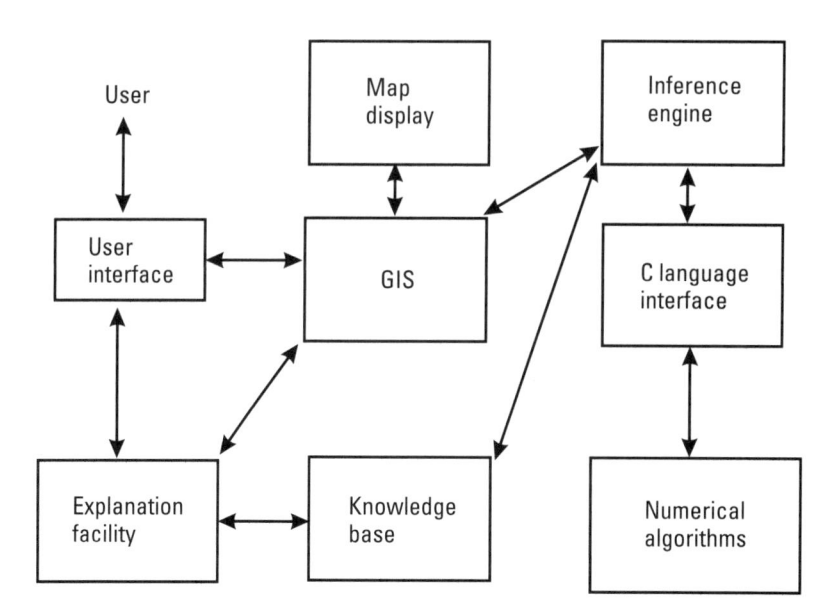

Figure 3.3 Software components of WAIMSS [1, 2].

implement some of the computationally intensive modules and to connect the distinct packages to one another. The aim of this section is to describe the relationship between the modeling and software development and present some of the important features of WAIMSS and the software packages used to develop WAIMSS.

The use and integration of different commercially available software products to develop different functions of WAIMSS, including database management, knowledge processing and inferencing, graphical and procedural program development, and GUI development are discussed in detail in Section 3.3.1.

3.3.1 Software Implementation Architecture

The WAIMSS architecture is implemented over a *server-centric* client-server architecture [1, 2]. The system uses a Sun Sparc 1000 as a central server with Pentium PCs running an X-Windows emulation software at the client sites. The idea is to have the central server at the TMC with agency-specific applications and databases developed for each client, as shown in Figure 3.4. In addition, the server also supports the blackboard and the TMC applications and databases. Client sites request data from the blackboard and view events related to the progress of the incident response on the blackboard. Communication among the different client applications and the blackboard was achieved using interapplication communication (IAC) in Arc/Info. An alternative approach would be to implement the same system to a *client-centric* environment to reduce the processing loads on the main server. Under that scenario, agency applications reside at their respective client sites.

The control mechanism for the server-centric system is implemented as a part of the TMC application using NEXPERT rules and Arc Macro Language (AML) macros. The NEXPERT rules are used to monitor the changes in the

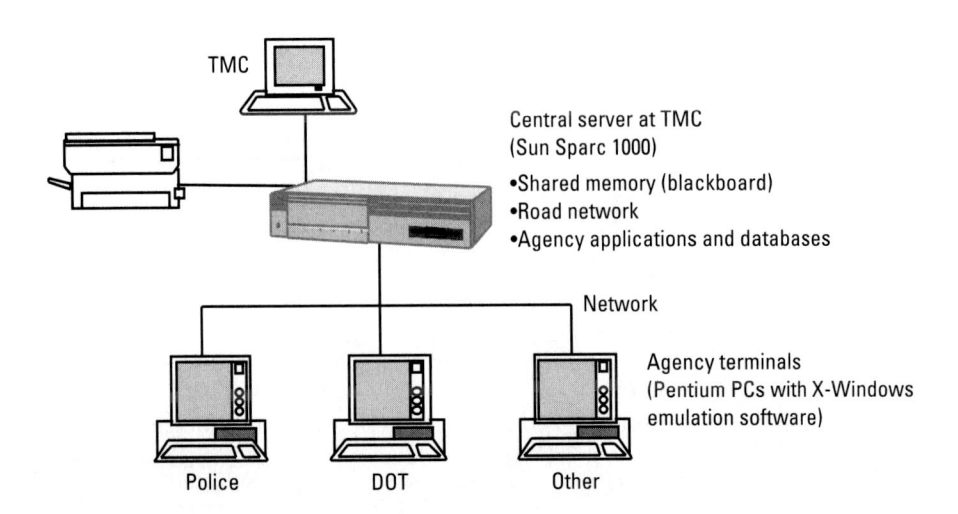

Figure 3.4 Integrated implementation architecture of WAIMSS [1, 2].

response plan and to determine when to broadcast messages and update displays at agency terminals. The AML macros are used to define roles and the necessary read/write restrictions for the various agencies. The read/write restrictions for the various agencies are an important feature of the implementation since they limit each agency's access to the appropriate part of the central database based on some predefined rules. That, in turn, guarantees the reliability of the incident database by preventing unauthorized read/write access to certain fields. For example, tow truck operators are not allowed to change link characteristics. Network data can be changed only by the TOC personnel. That way, keeping track of changes to the database and managing the database effectively are made easier.

The messaging and display updates initiated by the control mechanism require an additional input/output channel at each agency application. That is made possible by starting up each agency application as an Arc/Info server session and storing details about its host name, program number, and version number in a "connect-file" as a part of the process ID list. The additional input/output channel is used to pop up message windows and update displays from another agency application.

The individual components of the blackboard shown in Figure 3.1 were implemented as follows:

- *Response plan information.* The response plan for an incident constitutes the list of agencies that are needed to respond to the incident and the resources they will need. In addition, the information on diversion routes and the people who need to be contacted at each agency are made available. Current versions of WAIMSS stores the data in Info at the shared memory location on the server. An Open Interface front end and the Arc Map display are used to view the data from each client site through IAC client requests or automated display updates through the control mechanism.

- *Prioritized action list.* The prioritized action list is maintained in the shared memory location as a text file. The actions in the list are created by the control mechanism based on the current deficiencies in the response plan.

- *Static and dynamic network information.* Static and dynamic information on network conditions is stored as a relational database in Info. The Info files are accessed by each client application. The control mechanism is used to enforce the required read/write restrictions into the database based on the roles assigned to each agency. AML macros are used to define the roles.

- *Resource information.* Resource information is stored as related Info files. Resource data that need to be shared are stored at the shared memory location. Agency-specific databases may contain additional information that other agencies may not need. As in the case with static and dynamic network information, read/write restrictions are enforced by the control mechanism. Additional resource information is added to the existing files when necessary from the local databases by the TMC using the messaging component.
- *Process lists.* Process lists are maintained as text files at the TMC. The event list maintains a log of all the changes made to the central database. It is implemented using an AML macro that writes to the event list each time a write operation is executed on the central database. In addition, a list of the process IDs of each agency Arc/Info session is also maintained. The IDs are used by the control mechanism for communicating and passing messages between the different sessions.

3.3.2 Application Development

Developed in a UNIX environment on a Sparc 1000 server, WAIMSS applications are specially designed to combine the rule-based reasoning capabilities of the Nexpert-Object expert system shell with the spatial data-handling capabilities of Arc/Info [1, 2]. Linking an expert system with a GIS enables both the expert system and the GIS to perform new tasks and opens the way for more complex spatial analysis and more flexible querying of GIS databases. Figure 3.5 shows the overall design of the application, which was determined after careful consideration of the available GIS software and expert system shells. Arc/Info, developed and supported by the Environmental Systems Research Institute (ESRI) [7], is a powerful toolbox that supports the entire spectrum of GIS applications. Spatial objects in Arc/Info are represented as points, lines, or polygons. The locational data associated with those objects are defined through a topological model, while the thematic data are defined using a relational data model in Info. The basic unit of storage in Arc/Info is called a coverage and corresponds to a single layer of a map that contains information about one type of locational feature.

Arc/Info has a layered architecture, the foundation of which is the data engine used to access and manage the geographic database. At the next level, Arc/Info contains a powerful and flexible command language that provides access to sophisticated geoprocessing tools operating on the various data sources supported by Arc/Info. AML provides the developmental environment in which sophisticated macro procedures can be automated and custom user

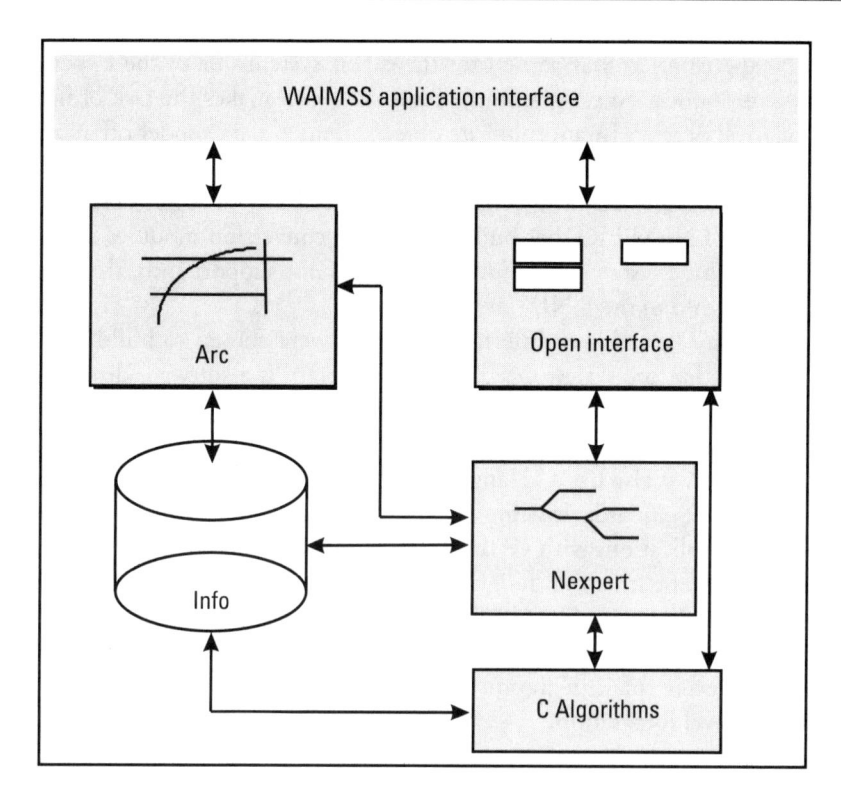

Figure 3.5 WAIMSS application design [1, 2].

interfaces can be built. A third method for accessing Arc/Info is through the use of IAC. IAC tools in AML permit other applications software to execute operations in Arc/Info, thus allowing its use as a GIS data and process server. Those capabilities of Arc/Info make it especially suitable for the design of the system described.

Nexpert-Object, developed and supported by Neuron Data, provides a framework for the construction of rules and an object-oriented model for representation of the data on which the rules act [8]. A NEXPERT rule consists of three parts: a series of conditions, a hypothesis, and a series of actions. If all conditions are true, then a hypothesis is true and can be used to execute a set of actions. NEXPERT's rules act on objects that are members of classes and that possess properties. Objects and classes are incorporated into the conditions and actions of the rules. Rules can change object class memberships of the values of their properties. Together the rule and the object networks form the knowledge base of the expert system.

Nexpert-Object was chosen for the expert system side of the Expert-GIS because its application programming interface (API) makes the task of integration relatively easier. In addition, its object-oriented data model offers a great deal of flexibility in data representation, and its ability to reason by both forward and backward chaining offers a similar flexibility in logic representation [9]. Although NEXPERT has built-in database conversion modules for several popular database systems, it unfortunately does not support Info, the database used by Arc/Info in the UNIX environment.

Neuron Data's Open Interface Elements were chosen to build the additional GUIs that the system needs. Open Interface is a software development environment that permits the development of cross-platform applications with native graphical interfaces. Open Interface comprises a set of libraries and a resource editor. It also has a C language API, which is a highly modular American National Standards Institute C library. The API provides facilities for developing applications with GUIs for any standard windowing environment.

A key element in the design of the application is bridging the different software environments of Arc/Info, Nexpert-Object, and Open Interface to permit easy transfer of data and command-level control between the software. The two levels of integration required are data-level integration and command-level integration.

3.3.3 Data-level Integration

An easy approach for data-level integration is to employ data files in a format that both Arc/Info and Nexpert-Object can support [1, 2]. However, this would require writing data into a common file format before it can be accessed by either software, making the operation a time-consuming and not suitable for real-time systems.

To allow for more direct data bridging, a C environment was implemented to permit direct access to the Info file structure and Nexpert-Object variables. The Nexpert-Object expert system development package includes an API, permitting developers to access NEXPERT variables from C programs. Open Interface, the GUI development kit, comes with a toolkit of C libraries that can be used to read and write widget attributes. Shareware C functions developed by Todd Stellhorn [7, 10] are used to allow NEXPERT to directly read network information from the Info files. That eliminates the otherwise tedious task of writing to a common file format.

3.3.4 Command-level Integration

A command-level interface permits Nexpert-Object and Arc/Info to issue commands to each other and act on the command issued by the other program [1, 2]. Although both packages permit the execution of external programs through system calls, only a command interface can provide for the transfer of control between the two programs. That level of integration is achieved using Arc/Info's IAC and by building an environment in C for controlling knowledge processing in NEXPERT. The API for NEXPERT makes it possible to build a controller for the inferencing and knowledge processing in C, which could suggest parameter values and hypothesis directly into the inference engine.

IAC in Arc/Info enables software applications on remote or local machines to communicate with each other [7, 10]. Based on Open Network Computing's (ONC) Remote Procedure Call (RPC) protocol, IAC provides a way for external applications to request services of an Arc/Info process in server mode (AI-Server) and also makes it possible for an AML application to exploit the capabilities of other applications by being a client of those applications. Important components of IAC are the AI-Server and AI-Client shells. The AI-Server provides a shell for creating a server to execute C functions and to have those functions directly accessible from Arc/Info's AML-IAC interface. That functionality is used to call NEXPERT executables with AML variables as arguments. AI-Client, on the other hand, provides the ability to create client applications in C that can execute Arc/Info commands. It was used to write C functions that Nexpert-Object could access and execute Arc/Info commands by connecting to an Arc/Info server process and passing requests to that process. A working version of WAIMSS can be obtained from Virginia Tech Center for Transportation Research. However, valid licences for NEXPERT and ARC-INFO are needed to be able to run the WAIMSS.

3.4 Summary

Proposed WAIMSS architecture combines the powerful spatial data-handling capabilities of a GIS with the rule-based logic of an expert system in a fully integrated Expert-GIS framework to provide interactive content and group process support for incident management operations. The implementation framework is based on a blackboard architecture to support incident management operations by providing facilities for spatial and temporal data analysis and a mechanism for interaction between different responding agencies. Because incident management is a multiagency effort, the client-server architecture of WAIMSS facilitates the incident response agencies in group decision making by providing data-sharing and data-exchanging capability. The use of specific commercial software packages to implement various functions of the proposed architecture proved to be an efficient and economical way of implementing that complex architecture. The data-level and command-level integration of those software packages is one of the most challenging aspects of the implementation process. Finally, it is important to emphasize the importance of a client-server architecture to effectively address the multiagency implementation requirements of a modern wide-area incident management decision support system.

Review Questions

1. Identify network, incident, and response data that should be accessed by each agency involved in incident management in your area.

2. Identify real-time data needs for a system similar to the one discussed in this chapter. Discuss the reasons for not allowing certain agencies to read/write certain types of data. Use the incident management handbook of your state.

3. Identify other GIS and expert system software packages that can be used to develop a similar system.

References

[1] Ozbay, K., A. Narayanan, and S. Jonnalagadda, "Wide-Area Incident Management Support System (WAIMSS) Software," *Proc. Third Annual World Congress on ITS*, Orlando, FLA, October 14–18, 1996.

[2] Kachroo, P., K. Ozbay, and W. Wu, "Development of a Wide-Area Incident Management System," (work order # DTFH71-DP86-VA-20), FHWA Final Report, 1997.

[3] Teng T. C. J., and K. Ramamurthy, "Group Decision Support Systems: Clarifying the Concept and Establishing a Functional Taxonomy," *INFOR*, Vol. 31, No. 3, 1993.

[4] Virginia Department of Transportation (VDOT), *Northern Virginia Freeway Management Team Operating Manual—A Regional Plan for Traffic Management on Northern Virginia Freeways*, April 1990.

[5] Ozbay K., A. G. Hobeika, S. Subramaniam, and V. Krishnaswamy, "A Heuristic Network Generator for Traffic Diversion During Non-Recurrent Congestion," *TRB Annual Conf.*, Washington, D.C., 1994.

[6] www.esri.com/industries/transport/transport.html, 1998.

[7] Environmental Systems Research Institute (ESRI), Inc., *Arc/Info User's Guide, Getting Started—Introduction to Arc/Info*, 1994.

[8] www.neurondata.com, 1998.

[9] Evans, T. A., D. Djokic, and D. R. Maidment, "Development and Application of Expert Geographic Information System," *J. Computing in Civil Engineering*, Vol. 3, 1993.

[10] Environmental Systems Research Institute (ESRI), Inc., *Arc/Info User's Guide, Customizing Arc/Info—AML (Arc Macro Language)*, 1994.

Selected Bibliography

Jonnalaggada, S., "An Expert-GIS System for Freeway Incident Management," Master's Thesis, Virginia Ploytechnic Institute and State University, Blacksburg, VA, 1996.

4
Incident Detection

4.1 Introduction

Incident detection and verification is the first step of an efficient incident management process. (It is a misnomer to call the first step of incident management process "incident detection" instead of the more complete term "incident detection and verification." However, it is well understood that incident verification is an integral part of this first step.) Incident detection can be defined as the process of identifying the spatial and temporal coordinates of an incident. It is important to emphasize verification as part of a complete incident detection concept, because most of the subsequent incident management steps are undertaken only after the existence of the incident has been verified.

This chapter attempts to give a satisfactory description of the state of the practice in the area of incident detection and verification. Because of the large body of literature that exists in this area, it is impossible to cover all possible algorithms. Our aim is to introduce the concept of incident detection and verification by presenting some of the more prominent algorithms that are widely used today. More important, some of the new field operation tests involving the deployment and evaluation of incident detection techniques are presented. By doing so, we hope to give the reader a snapshot of recent and interesting incident detection projects across the United States.

Another important aspect of this chapter is its presentation of developments in the area of surface street incident detection rather than limiting the discussion to freeway incident detection. The latter usually receives more attention for many reasons, which are discussed later in this chapter.

4.2 What Is Incident Detection?

Incident detection is not a new concept in traffic management. It has been around since the mid-1960s and early 1970s as part of standard traffic/incident management practice. In the 1960s, loop detector occupancies greater than 40% were used in Chicago to identify high reduction in capacity and as an early warning of potential freeway incidents. Incident detection, which is the process of determining the presence and location of an incident, has two major steps:

1. Determine the existence of congestion using data obtained from a surveillance system.
2. Analyze the data to determine if the cause of the congestion is an incident.

4.2.1 Traffic Surveillance and Data

The first step requires some kind of traffic surveillance system, which is also an integral part of the traffic management process. In traffic engineering terminology, *surveillance* denotes the real-time observation of traffic flow conditions in time both and space. The first step of incident detection process, namely, surveillance of roads, has been around since the early 1960s. Detroit was the first U.S. city to implement a surveillance system that comprised CCTV cameras, loop detectors for traffic detection, and variable signs for control purposes. The system became operational in the spring of 1961. Since then, almost all the major cities in the United States implemented a form of traffic surveillance for traffic monitoring and control purposes. The traffic surveillance systems include a variety of traffic sensors on the road and highway safety or police patrols. In addition to those data sources, drivers report incidents using emergency call boxes, cellular phones, and citizens band (CB) radio. The TOC plays an important role by coordinating the incoming data regarding incidents and making decisions for managing the incidents. Thus, the surveillance systems coupled with other data sources and the TOC play an important role in incident detection.

4.2.2 Analysis of Traffic Data

The increasing number of incidents and the subsequent need for quick detection to reduce adverse effects of incidents resulted in the development of automated incident detection (AID) systems. The analysis of traffic data is done using AID algorithms developed over the last three decades. Courage and Levin developed the earliest AID algorithm [1], which was implemented on the John C. Lodge Freeway in Detroit.

4.2.3 Importance of Incident Detection Time

Incident detection can be seen as a crucial component of the overall incident management process. It is clear that an incident has to be detected and verified before any other incident management actions can be taken. To guarantee the success of any incident management process, it is critical that incidents are detected as soon as they have occurred. Timely and accurate incident management becomes more important when we consider the negative effects of not clearing an incident as quickly as possible. A delay in detecting an incident can cause long queues and traffic congestion, which, in turn, are the primary cause of secondary accidents.

Another important advantage of quick incident detection is the possible reduction of fatalities. In that context, *fatality* is defined as a person involved in a motor vehicle crash dying within 30 days of the accident. In a recent report [2], Evanco states that the outcomes associated with injury trauma are time dependent. He reports that the initial results of ITS operational tests indicate that if more effective incident detection techniques are used the incident detection and verification time (accident notification time) can be reduced to 2–3 minutes, from an estimated average of 5.2 minutes in 1990.

In the same report, Evanco develops a Poisson regression that estimates the number of fatalities as a function of variables such as, vehicle miles traveled (VMT), mean vehicle speed (MVS), alcohol consumption per capita (ACC), young/aged driver (YAD) fraction, accident notification time (ANT), and personal income per capita (IPC). Evanco uses the 1990 data from individual states in the United States to calibrate the following relationship, which is expressed in logarithmic:

$$\ln(NF_i) = a_0 + a_1 \cdot (VMT_i) + a_2 \cdot (MVS_i) + a_3 \cdot (ACC_1) + a_4 \cdot (YAD_i) + a_5 \cdot (ANT_i) + a_6 \cdot (IPC_i) \quad (4.1)$$

In (4.1), the subscript *i* represents the state *i* in the United States; ln () is the natural logarithm; and a_0, a_1, a_2, a_3, a_4, a_5, and a_6 are the model parameters that need to be estimated using actual incident data. It is important to note that Evanco's study focuses on urban freeways. The data set he used contained a total of 2,331 fatalities on 11,500 miles of urban freeways. Thus, his numbers regarding the reduction of fatalities reflect the total number of fatalities on urban freeways rather than the total number of highway fatalities in the United States.

Equation (4.1) can be used to determine the effect of a change in the notification time, D(*ANT*), on a change in the number of fatalities, D(*NF*), as follows:

$$\frac{\Delta NF}{NF} = 0.27 \cdot \frac{\Delta ANT}{ANT} \qquad (4.2)$$

Using the relationship shown in (4.2), Evanco concludes that if the 5.2-minute accident notification time is reduced to 3 minutes through the introduction of a more effective incident detection program, there would be an 11% reduction in the number of fatalities nationally. That would translate into 246 fewer fatalities in a year.

Using an estimated monetary cost of $111,870 for a nonfatal injury and $708,235 for a fatality and an estimated comprehensive cost of $560,018 for an injury and $2,074,533 for a fatality, he calculated the net benefit of avoiding fatalities. The comprehensive cost takes into account the additional cost due to the loss of the quality of life. By reducing the notification time from the average 5.2 minutes to 3 minutes on both urban interstates and rural freeways, the net monetary benefit is estimated to be $267,767,900 and the net comprehensive benefit to be $931,465,300.

It is important to understand that Evanco's study deals exclusively with the reduction of detection times, not the whole incident duration. However, the results of this timely research project clearly demonstrate the importance of rapid incident detection in terms of saving lives as well as reducing the costs associated with traffic accidents. It is clear that rapid incident detection not only will reduce congestion due to incidents but will also reduce the number of fatalities. The possible improvement in public safety that can be achieved by an effective incident management program is an important benefit to any DOT trying to determine the benefits and the costs of a new or an existing incident management program.

4.3 Effect of Incident Detection Time on Overall Incident Duration

As mentioned in Chapter 3, overall incident duration time has several components, including incident detection and verification time, incident response time, and time to normal flow. Thus, it is important to have answers to the following questions:

- Is the incident clearance time affected by the length of incident detection time or type of detection source?
- Is the incident verification plus incident response time affected by the incident detection time and or source?

If the answer to the first question is "Yes," it is clear that by reducing incident detection time we can also reduce the incident clearance time. Also, if the detection source can be shown to have a major effect on the incident clearance time, future investments can be channeled to that specific kind of detection source. The same arguments are valid for the second question.

A recent report [3] studied the effect of incident detection on the overall and individual components of incident duration using the I-880 incident database. The I-880 incident database was developed specifically for the California Freeway Service Patrol (FSP) study [4]. The incident data were recorded by probe vehicles traveling on the 9.2-mile section of the I-880 freeway in the city of Hayward, Alameda County. The database was complemented by the California Highway Patrol (CHP) computer-aided dispatch (CAD) logs. For the development of the incident duration models, Skabardonis and his colleagues at PATH at the Institute of Transportation Studies at the University of California, Berkeley, employed the CAD database. Their study tested several relationships that can answer the two questions posed at the beginning of this section. First, it was shown that incident clearance time is not significantly affected by the incident detection time or the incident detection source, but it is mainly affected by the incident type. Then the effect of incident detection time on the incident verification plus incident response time was tested. Incident detection time was found to be an *insignificant independent* variable for the incident verification plus response time. Further tests showed that incident verification plus response time are clearly affected by the incident type and incident detection source.

Those results show that the answer to both questions is "No." The length of incident detection time is not a significant factor that affects incident response and clearance times. However, it was also observed that the detection source affects the verification and response times. Incidents detected as a result of a cellular phone call require longer times to verify and respond to compared to the incidents detected by FSP or CHP. That is obvious because an FSP or CHP officer who detects an incident is at the scene and ready to take immediate action, whereas a CHP officer has to be dispatched to the scene to verify an incident reported over a cellular phone.

66 Incident Management in Intelligent Transportation Systems

Although incident detection and verification times do not have a direct effect on the response and clearance time, it is important to understand that the shorter the incident detection and verification time, the shorter the small incident duration (See Figure 1.2).

4.4 Incident Detection Issues

There are three basic issues concerning incident detection: surveillance issues, algorithmic issues, and verification issues. Each of these issues and its relevance to the incident detection process are discussed in detail in this section.

4.4.1 Surveillance Issues

Traffic surveillance can be defined as the process of measuring traffic flow characteristics and sending this information to the TOC. The traffic surveillance system is the main source of the traffic flow data employed by almost all the AID algorithms. The last decade has been an exciting time as far as traffic surveillance is concerned. Since the inception of ITS, many new and effective traffic sensors have been introduced. Although loop detectors are the most widely used sensing system, several new sensors using different technologies have been widely adopted by DOTs throughout the country. The following detector technologies cited in [5] are currently used in traffic sensors.

- Inductive loop technology is an active detector technology that responds to ferrous mass (cars).
- Magnetometer technology is a passive detector technology that also responds to ferrous mass (cars).
- Infrared technology, which can be either active or passive. Passive infrared technology uses the contract in thermal radiation to detect vehicles, while active infrared technology makes use of reflected signals.
- Acoustic detection technology is passive and employs sound to detect vehicles.
- Ultrasonic detection technology is active and employs reflected sound to detect vehicles.
- Charged coupled devices (CCD) camera is a passive technology that uses the contrast in visible light.

- Doppler radar detection is an active detection technology that uses frequency shift of reflected signal.
- Pulsed radar technology is an active technology that makes use of reflected signals.

Although there is a large range of detector technologies, all sensors are evaluated according to the quality of surveillance data they provide. The factors considered to evaluate them are:

- Reliability;
- Performance under different environmental conditions;
- Data accuracy;
- Real-time performance.

In addition to those four factors, cost plays a pivotal role in the selection of any detector system. It is clear that a detector that provides excellent results in terms of the four factors will be highly undesirable if it is very expensive compared to the other available technologies.

Reliability of a sensor is most of the time on the top of the list of evaluation factors. Many traffic surveillance systems are plagued by the low reliability of loop detectors. Frequent malfunctions or breakdowns of sensors can seriously hamper the performance of the overall traffic management operations. It is often very expensive and impractical to fix or replace a sensor that is placed under the road surface.

Performance of a specific sensor under different environmental conditions is an important factor. Among other things, a study conducted by Hughes Aircraft [5] determined the best environmental conditions for different sensor technologies, as listed in Table 4.1. The table clearly shows that different sensors using different technologies can perform better under certain environmental conditions such as fog, rain, or smoke. A simple example is the inductive loop technology, which does not perform well when snow covers the road surface. It is important to choose the sensor that will perform well under the specific environmental conditions of the area where it will be deployed.

Data accuracy is another important factor that largely depends on the installation and calibration of the sensor. A sensor that is known to be very accurate can collect erratic data if it is not appropriately installed and calibrated. Another important issue is continuous monitoring of the performance of the sensor to ensure its accuracy, because over time sensors need to be

Table 4.1
Best Environmental Conditions for Various Sensor Technologies
(*From:* [5], with permission for Elsevier Science, ©1998.)

Technology	Clear Day	Clear Night	Clear Night	Hot Day	Light Wind	High Wind	Light Rain	Hard Rain	Light Snow	Hard Snow	Fog	Smoke	Weather Monitor
Inductive Loop	✓	✓	✓	✓	✓	✓	✓	✓			✓	✓	
Magnetometer	✓	✓	✓	✓	✓	✓	✓	✓			✓	✓	
Infrared (passive)	✓	✓	✓	✓	✓	✓					✓		
Infrared (active)	✓	✓	✓	✓	✓	✓					✓		
Acoustic (passive)	✓	✓							✓	✓	✓		
Ultrasonic	✓	✓			✓		✓		✓				
CCD camera	✓	✓	✓	✓	✓	✓						✓	
Radar–Doppler	✓	✓	✓	✓	✓	✓					✓		
Radar–FMCW	✓	✓	✓	✓	✓	✓					✓		
Laser–Pulsed	✓	✓	✓	✓	✓	✓							

(From R. Weill et al., with permission from Elsevier Science, 1998)

recalibrated. More important, accuracy of sensor evaluations should be made for the operational conditions for which they are designed. Some sensors are designed for low-volume roads. Therefore, testing those sensors under high-volume traffic conditions will produce unacceptable accuracy results. However, that does not mean the sensor is not accurate. It simply means the sensor is not tested under right conditions.

Finally, real-time performance of any sensor plays a crucial role in ensuring timely decisions. In the past, real-time performance was not an issue because the data were used for mostly off-line purposes. Today, the focus is on on-line applications, and sensors are expected to collect second-by-second data. Of course, that raises the issue of real-time performance, which easily can be addressed by the use of widely available high-speed communication technologies.

The selection of a specific sensor technology also affects the type of AID algorithms that can be used and vice versa. Different sensors can measure different traffic variables, and different incident detection algorithms need

Table 4.2
Traffic Flow Parameters Measured or Calculated by Various Sensors [6]
(*From:* [5], with permisison of Elsevier Science, ©1998)

Technology	Volume	Speed	Class	Occupancy	Density	Headway
Inductive Loop	✓	✓	✓	✓		✓
Magnetometer	✓	✓	✓	✓		
Infrared (passive)	✓	✓		✓	✓	
Infrared (active)	✓	✓				
Acoustic (passive)	✓			✓		
Ultrasonic	✓	✓	✓	✓	✓	✓
CCD camera	✓	✓	✓	✓		
Radar-Doppler	✓	✓	✓	✓		
Radar-FMCW	✓	✓	✓	✓		
Laser-Pulsed	✓	✓	✓	✓	✓	

different data types. Table 4.2 shows traffic flow parameters measured or calculated by various sensor technologies [5]. Thus, extreme care must be taken in the selection of the specific sensor technology of incident detection algorithm and making sure that they are compatible.

4.4.2 Algorithmic Issues

This section focuses on two types of algorithms used for automated incident detection on freeways. Until recently, most of the automated incident detection algorithms were based on traffic flow measurements made at one point. This family of algorithms is generally called *point-based algorithms*. The point measurement–based algorithms use the following approaches to detect incidents on the freeways:

- Comparative or pattern recognition;
- Statistics;
- Traffic model and theoretical algorithms;
- Artificial intelligence–based algorithms.

The other family of incident detection algorithms is called *spatial measurement–based algorithms*. They make use of video cameras and image-processing techniques, which are becoming more common in traffic-engineering applications.

The discussion in this section based on a comprehensive review presented in [6] and [7] and related papers [8-22] focuses more on the first family of algorithms, because they are by far the more common types of applications and they employ already existing and relatively cheap sensor technologies, such as loop detectors.

4.4.2.1 Comparative or Pattern Recognition–Based AID Algorithms

Comparative AID algorithms, that is, algorithms based on pattern recognition, were the first ones to be developed by traffic engineers. They rely on recognizing and differentiating unusual patterns of traffic from "normal" traffic conditions. This type of algorithm looks at the occupancy levels at the detector stations upstream and downstream of the incident. The basic principle of these algorithms is that an incident will create increased occupancy levels upstream of the incident and a decrease downstream. The measured values of the traffic flow are compared with predetermined threshold values using a decision tree logic. The algorithm determines the existence of an incident if the threshold values are exceeded. "California algorithms," developed as a result of a 1973 FHWA-sponsored study, are the best known comparative/pattern recognition algorithms. The study produced 10 versions of the California algorithm. A decision tree that depicts the general logic of the California algorithm is shown in Figure 4.1. In that decision tree, measured traffic flow variables (occupancies) pass three sequential steps. At each step, the occupancies are compared with a predetermined threshold value. The first step compares the occupancy difference between the downstream and upstream stations with the threshold value. The next two steps look at the relative spatial and temporal differences of occupancies. An incident is declared if all three threshold values are exceeded.

Among those 10 California algorithms developed by [8], algorithms #7 and #8 are known to produce best results [7]. The decision tree for California algorithm #7 is shown in Figure 4.2.

4.4.2.2 Statistical AID algorithms

Statistical algorithms model the stochastic traffic flow patterns obtained from the loop detector data. Among the several models developed using statistical principles are as follows:

- The Standard Normal Deviation (SND), developed at the Texas Transportation Institute (TTI) by [9], is based on the assumption that

Figure 4.1 Basic California algorithm [8].

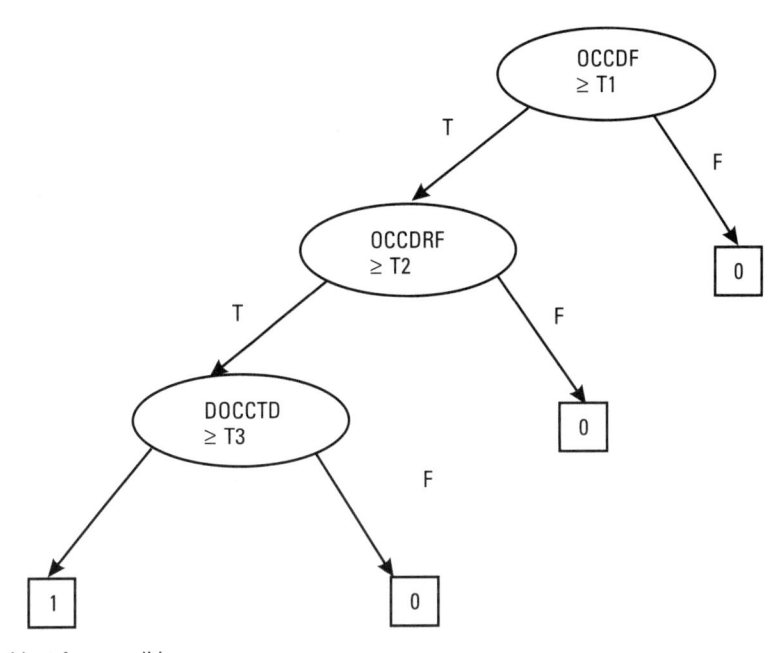

0 = Incident-free condition
1 = Incident condition

Typical threshold values
T1 = 8
T2 = .5
T3 = 0.16
OCCDF = Spatial difference in occupancies
OCCDRF = Relative spatial difference in occupancies
DOCTD = Downstream occupancy

a sudden change in traffic flow pattern due to an incident is the sign of the occurrence of an incident. In this algorithm, the SND is defined as the number of deviations away from the mean. The algorithm compares occupancy averaged over 1-minute intervals with the historical values of the mean and the SND. If the SND exceeds a critical value, an alarm that signals the occurrence of an incident is triggered.

- The Bayesian algorithm, developed by Levin et al. [10], uses Bayesian statistical techniques and historical data to determine the probability of an incident signal being caused by downstream lane blocking [10]. It calculates the probability that the relative occupancy differences between detectors are caused by an incident.

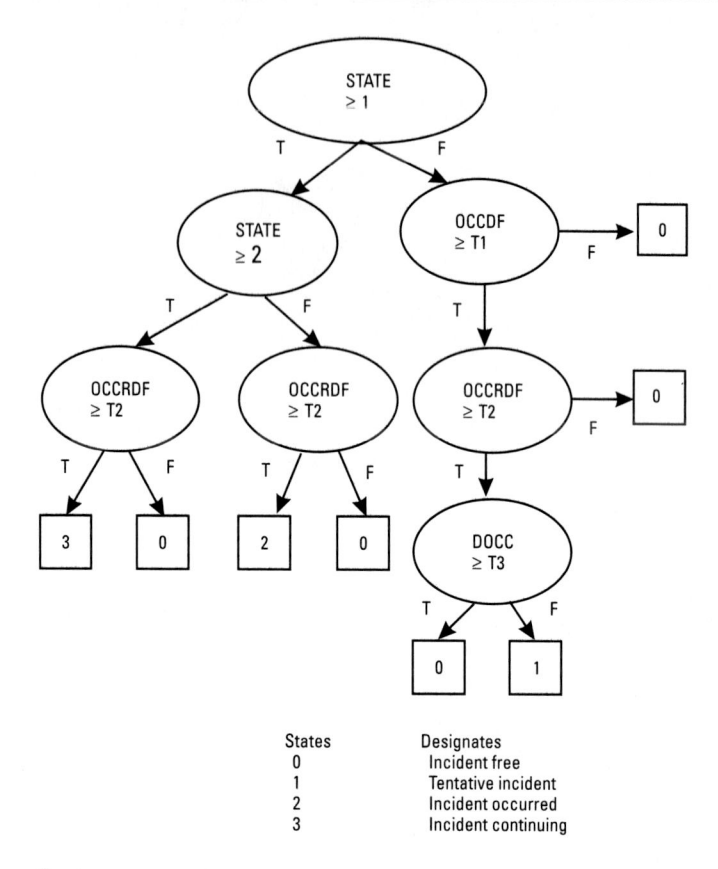

Figure 4.2 Decision tree for California algorithm #7 [8].

States | Designates
0 | Incident free
1 | Tentative incident
2 | Incident occurred
3 | Incident continuing

- The time series and filtering algorithms developed by [11–15] compare short-term predictions of traffic conditions to measured traffic conditions.

4.4.2.3 Traffic (Modeling) and Theoretical Algorithms

Theoretical algorithms actually employ the basic theories of traffic flow characteristics. Among the most notable of these types of AID algorithms is the McMaster algorithm, which makes use of the catastrophe theory to model the sharp changes in traffic flow. Use of the catastrophe theory was first proposed by [16, 17]. The algorithm develops a volume-occupancy template divided into four distinct areas, each corresponding to a particular state of traffic flow (Figure 4.3). As shown in Figure 4.3, state 1 represents uncongested traffic flow conditions, states 2 and 3 represent areas where congested flow conditions

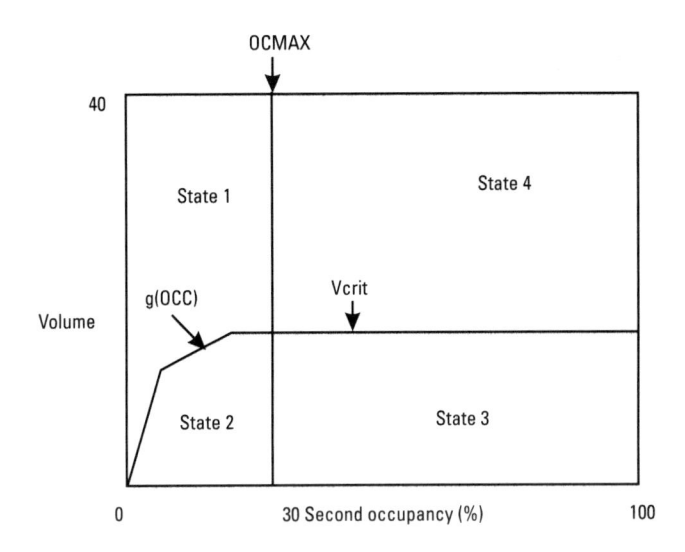

Incident Detection 73

Figure 4.3 Illustration of volume-occupancy template for traffic state classification. (*From*: [17], with permisison from the Transportation Research Board and the author.)

Volume = Number of vehicles in 30 seconds
g(OCC) = K*b*occupancy, 0 < k < 1, 0, (*k, a, b*, and *V*crit are station-specific parameters.)
OCMAX = Maximum uncongested occupancy

occur, and state 4 represents traffic flow conditions at the downstream of a permanent bottleneck. The algorithm makes two basic comparisons between this figure and actual loop data. The first test determines if the actual detector is congested. If congestion exists, the algorithm tries to determine the source of congestion by examining the traffic state at a downstream detector. The McMaster algorithm is rated as one of the most accurate incident detection algorithms with an overall detection accuracy between 70% and 85% and a false alarm rate of 1% or less [7].

4.4.2.4 Artificial Intelligence–Based Algorithms

Some researchers use artificial neural networks (ANNs) to develop models for AID. An ANN consists of simple processing elements and neurons that are interconnected. The concept of ANNs is borrowed from the human brain. The idea is to train the ANN by feeding it with input and associated output data. As a result of that training process, ANN develops rules of associations among its neurons. For incident detection, input data consist of traffic flow variables such as volume, speed, and occupancy at both upstream and downstream detectors. Some researchers who have used ANNs for AID have reported successful results [18, 19].

4.4.2.5 Surface Street Incident Detection (SSID)

Surface street incident detection still remains one of the biggest challenges in the area of AID. The interrupted flow conditions due to the traffic signals create extra difficulties for the development of reliable AID algorithms. Some of the basic problems associated with surface street incident detection can be summarized as follows:

- Interrupted traffic flow conditions complicate the analysis of traffic flow data;
- Arterials are much longer than freeways and require more personnel and equipment for incident detection and verification;
- Most arterials are not instrumented with traffic detectors, which makes it impossible to use AID algorithms.

Several researchers [20–22] have recently attempted to develop incident detection systems for arterials. Among the most notable ones is the work presented by Bhandari et al. [20]. That project developed an arterial incident detection system for the ADVANCE project, an advanced traveler information system demonstration in the northwest suburbs of Chicago. To remedy some of the problems mentioned above, they used data from various sources, including fixed detectors, probe vehicles, and anecdotal sources. All the data was processed through the use of data fusion algorithms, and the likelihood of the occurrence of an incident at a given location was determined. One of the most important aspects of the algorithm is the use of anecdotal data, which basically are descriptions of incidents reported by trained and untrained field observers. Several other SSID algorithms developed as part of the ADVANCE project were using data fusion approaches. They produced incident detection rates that were between 66.7% and 87.0% [23]. A good review of the state of the SSID algorithms and research needs is given in [24].

4.5 Verification Issues: Evaluation of Incident Detection Systems

Three basic measures of effectiveness (MOEs) are used to evaluate incident detection algorithms:

- The detection rate (%) is the ratio of the number of detected incidents to the actual number of incidents in the data set.

- The false alarm rate (%) is the fraction of incorrect detections to the total number of algorithm applications.
- The time to detection (sec) is the average time required for the algorithm to detect an incident.

Those MOEs are not independent. For example, longer time to detection means lower false alarm rates. However, it is clear that that is a tradeoff between loosing precision time and lowering false alarm rates. Another problem is the lack of standards in the evaluation methodology of AID algorithms; hence, the evaluation results are not comparable most of the time. A summary of the reported performance of various algorithms is given in University of California PATH ITS LEAP Web page [6]. The evaluation results of AID algorithms here are shown in Table 4.3. It is important to emphasize once more that the evaluation results are not obtained using a standard methodology. Thus, it may be misleading to compare the results with one another without considering the differences in the actual traffic and network conditions.

Table 4.3
Summary of the Evaluation Results of the Most Commonly Used AID Algorithms
(*From:* [6], with permission from UC Berkeley PATH ATMIS Group)

Algorithm Type	Detection Rate (%)	False Alarm Rate (%)	Average Detection Time (secs)
Basic California	82	1.73	0.85
California #7	67	0.134	2.91
California #8	68	0.177	3.04
Standard Normal Deviate	92	1.3	1.1
Bayesian	100	0	3.9
Time Series (Autoregressive Integrated Moving Average)	100	1.5	0.4
Exponential Smoothing	92	1.87	0.7
Low-Pass Filter	80	0.3	4.0
Modified McMaster	68	0.0018	2.2
Multi Layer Feed Forward Neural Networks	89	0.01	0.96
Probabilistic Neural Networks	89	0.012	0.9
Fuzzy Sets	Good	Good	Up to 3 minutes quicker than conventional algorithms

4.6 Operational Field Tests

In addition to the conventional algorithms, there have been some recent attempts to use and evaluate more innovative and practical approaches for incident detection. Two of the most interesting and successful of these incident detection operational field tests are:

- The TRANSCOM TRANSMIT project implemented along a corridor in Staten Island, New York, centered in Interstate 287 and extending from the Verrazano Bridge across the New Jersey Turnpike to Routes 1 and 9 in New Jersey;
- The I-880 field experiment to evaluate the effectiveness of incident detection using cellular phones conducted on the 9.2-mile stretch of I-880 freeway in the city of Hayward, California (Alameda County).

4.6.1 TRANSCOM TRANSMIT Project

The TRANSMIT project was initiated to evaluate the feasibility of using electronic toll and traffic management (ETTM) equipment for traffic surveillance and incident detection purposes [20]. Toll tag readers installed along the roadway are also used to obtain vehicle surveillance data. The passage times of individual vehicles are kept track of by a central system which also calculated vehicle travel times by using this information. Incidents are detected by comparing actual vehicle travel times with historical vehicle travel times. If a predetermined number of vehicles are delayed at a certain location more than usual, an incident is identified. The system was evaluated by researchers from the New Jersey Institute of Technology (NJIT) in 1996. Table 4.4 summarizes the evaluation results [25].

In this study, between 9.8 and 15% operational false alarm rates were calculated by dividing the number of false alarms by the total number of alarms. Although that is a better measure of effectiveness for the evaluation of a specific AID algorithm, most of the time false alarm rates are calculated differently. Each AID system has a cycle of 5 to 10 seconds. At the end of each cycle, the system checks the existence of an incident. A cycle of seconds means 267,840 checks in a month (31 days). Thus, if an algorithm had 268 false alarms in a month, its false alarm rate would be 0.01. This gives a more comparable measure of the TRANSMIT system's performance with resppect to other AID systems using this in other places. When false alarm rates are calculated using this new method, they are found to be between 0.002% and 0.008%. However, even with such low numbers, the authors of [25] observed an average of 2

Table 4.4
Evaluation results of TRANSMIT.(*From*: [25] with permission from the author.)

Months	Valid Incidents	False Alarms	Total Alarms	Operational False Alarm Rates (%)	False Alarm Rates
January	67	6	73	8	0.002
February	113	20	133	15	0.008
March	158	14	172	8	0.005
April	140	19	159	12	0.007
Total	478	59	537	11	0.007

alarms for every three days, which might create operational problems for online applications. Finally, they concluded that operational false alarm rates gave a more realistic and better understanding of the system performance compared to extremely low false alarm rates which tend to mask the real performance of an AID program.

4.6.2 I-880 Field Experiment: Incident Detection Using Cellular Phones

The I-880 study attempted to evaluate the quality and adequacy of cellular phone information as part of advanced incident management systems. It also evaluated the effects of the timeliness of incident detection on incident duration and the effects of incident duration on congestion.

The study compared cellular phone incident data with incident data obtained from other detection sources. The major goal of the study was to assess the adequacy of cellular phones as an incident detection technique. Table 4.5 summarizes the comparison.

A close look at Table 4.5 shows that cellular phones have high false alarm rates. That is due basically to the reporting person's judgment of whether a vehicle is involved in an accident. However, in other parts of the country cellular phones are found to be an excellent information source for incident detection. Although false alarm rates will remain a problem, cellular phones can be considered one of the fastest and cheapest ways of detecting incidents. Coupled with other conventional incident detection techniques such as safety patrols, CCTV cameras, and loop detectors, the reliability of the cellular phone information can be drastically improved.

Table 4.5
Comparison of Incident Detection and False Alarm Rates (%) [3]
(© 1998, with permission from UC Berkeley PATH ATMIS Group)

Detection Source	Detection Rate		False Alarm Rate	
	Incidents	Other Events	Incidents	Other Events
Cellular Phone	37.9	1.2	7.4	32.0
California Highway Patrol (CHP)	25	4.3	0.0	0.0
Freeway Safety patrol(FSP)	17.1	4.9	5.4	0.0
Public Entity	13.3	0.6	0.0	11.1
Call Box	4.5	3.6	0.0	7.1

4.7 Summary

This chapter briefly discussed different incident detection techniques. As the first step of incident management process, incident detection is important in any successful incident management program. Timely and quick incident detection also has been shown to save lives and money [2]. The evaluation results of some of the most common incident detection algorithms also were presented in this chapter. Among them, the California algorithms and the McMaster algorithm are the AID algorithms most widely used by the DOTs because of their proven performance in accurately detecting incidents and their low false alarm rates.

An important issue in getting high performance from AID algorithms is the importance of proper calibration of threshold values. The use of new technologies such as cellular phones and ETTM for incident detection is part of the recent operation field tests conducted in the United States. Those new approaches for AID aim at taking full advantage of already existing technologies such as toll tags and cellular phones to monitor the traffic conditions. They appear promising in terms of developing effective AID systems for a relatively low investment on areawide surveillance infrastructure, such as traditional loop detectors or more expensive vision-based cameras. It is extremely important to emphasize the role and the effectiveness of freeway patrol programs for reducing the detection and verification times of incidents. Several recent studies [3, 4] clearly show that those new incident detection techniques are still not as good as freeway patrol programs such as FSP and CHP in California. Although AID traditionally has been used for freeway incident detection, there is a growing interest among researchers and practitioners for the development of AID algorithms for surface streets, too [24]. However, due to the lack of adequate surveillance on the surface streets and complex traffic environment compared to the freeways, surface street AID continues to be a challenging theoretical and practical problem for traffic engineers.

Review Questions

1. How would you determine the threshold values used in the California incident detection algorithm?

2. Choose two of the AID algorithms presented in this chapter and determine the surveillance needs for effectively using those algorithms.

3. Discuss the problems associated with calibrating and evaluating incident detection algorithms using traffic simulation packages.

4. Compare the two methods of false alarm rate calculations using a numerical example. Discuss the advantages and disadvantages of both methods.

5. Identify the AID techniques used in your state. Propose new AID techniques that can be used in your state and explain why.

References

[1] Courage, K. G., and M. Levin, *A Freeway Corridor Surveillance and Control System*, Texas Transportation Institute, Texas A&M University, College Park Station, Research Report 488-8, 1968.

[2] Evanco, M. W., *The Impact of Rapid incident Detection on Freeway Accident Fatalities*, Mitretek Report, McLean, VA, 1996.

[3] Skabardonis, A., T. Chira-Chavala, and D. Rydzewski, *The I-880 Field Experiment: Effectiveness of Field Experiment Using Cellular Phones,* California PATH Research Report, UCB-ITS-PRR-98-1 University of California, Berkeley, CA, 1998.

[4] Skabardonis, A., et al., *Freeway Service Patrol Evaluation*, PATH Research Report, UCB-ITS-PRR-95, Institute of Transportation Studies, University of California, Berkeley, 1995.

[5] Weill, R., J. Worton, and A. Garcia-Ortiz, "Traffic Incident Detection: Sensors and Algorithms," *Mathematical and Computer Modeling J.*, Vol. 27, No. 9–11, 1998, pp. 257–291.

[6] Black, J., *Automated Incident Detection Algorithms*, University of California, Berkeley, PATH LEAP Web page, 1998.

[7] U.S. Department of Transportation, "Freeway Management Handbook," Report No. FHWA-SA-97-064, August, 1997.

[8] Payne, H. J., E. D. Helfenbein, and H. C. Knobel, *Development and Testing of Incident Detection Algorithms: Volume 2—Research Methodology and Detailed Results,* Report No. FHWA-RD-76-20, Washington, D.C.: Federal Highway Administration, 1976.

[9] Dudek, C., C. J. Messer, and N. B. Nuckles, "Incident Detection on Urban Freeways," *Transportation Research Record* 495, Washington, D.C., 1974, pp. 12–24.

[10] Levin, G.M., Krause, 'Incident Detection: A Bayesian Approach," Transportation Research Record, TRB, Washingtin D.C. 1978, pp. 52–58.

[11] Ahmed, S. A, and A. R. Cook, "Analysis of Freeway Time-Series by Using Box Jenkins Techniques," *Transportation Research Record* 722, Washington, D.C., 1979, pp. 1–9.

[12] Masters, P. H., J. K. Lam, and K. Wang, "Incident Detection Algorithms for COMPASS—An Advanced Traffic Management System," *Proc. Vehicle Navigation and Information System Conf.*, Dearborn, MI, 1991, pp. 295–310.

[13] Cook, A. R., and D. E. Cleveland, "Detection of Freeway Capacity Reducing Incidents by Traffic Stream Measurements," *Transportation Research Record* 495, Washington, D.C., 1974, pp. 1–11.

[14] Stephanedes, Y. J., and A. P. Chassiakos, "A Low Pass for Incident Detection," Applications of Advanced Technologies in Transportation Engineering, *Proc. Second International Conf.*, Minneapolis, 1991, pp. 378–382.

[15] Willsky, A., et al., "Dynamic Model Based Techniques for the Detection of Incidents on Freeway," *IEEE Trans. Automatic Control*, Vol. AC-25, No. 3, 1980, pp. 347–359.

[16] Persaud, B. M., and F. L. Hall, "Catastrophe Theory and Patterns in 30 Second Freeway Traffic Data Implication for Incident Detection," *Transportation Research*, Vol. 23A, No. 2, 1989, pp. 103–113.

[17] Gall, A. I., and F. L. Hall, "Distinguishing Between Incident Congestion and Recurrent Congestion: A Proposed Logic,"1232, *Transportation Research Board, National Research Council ,Washington, D.C., 1989, pp. 1–8.*

[18] Cheu, R. L., et al., "Investigation of a Neural Network Model for Freeway Incident Detection," in B. H. V. Topping, ed., *Artificial Intelligence and Civil and Structural Engineering,* Civil-Comp Press, 1991, pp. 267–274.

[19] Ritchie, S. G., R. L. Cheu, and W. W. Wrecker, "Freeway Incident Detection Using Artificial Neural Networks," *Engineering Foundation Conf.*, Ventura, CA, 1992.

[20] Bhandari, N., et al., "Arterial Incident Detection Integrating Data From Multiple Sources." *Transportation Research Board,* Preprint # 950122, Washington, D.C., 1995.

[21] Cullip, M. J., and F. L. Hall, "Incident Detection on an Arterial Roadway," *Transportation Research Board*, Preprint # 970169, Washington, D.C., 1997.

[22] Ivan, N., et al., "Real-Time Data Fusion for Arterial Incident Using Neural Networks." *Transportation Research Board*, Washington, D.C., 1994.

[23] Iran, J.N., and V. Seth, "Data Fusion of Fixed Detector Probe Vehicle Data for Incident Detection," *Proc. of the Neural Network Applications in Highway and Vehicle Engineering Conference,* April, 1996, pp. 117–130.

[24] Franzese, O., and A. Rathi, *Proc. of the Workshop on Surface Street Incident Detection,* August 4–6, 1996, ITS Research Program, Scottsdale, AZ, Center for Transportation Analysis, Oak Ridge National Laboratory.[24]

[25] Mouskos, K. C., E. Niver, and S. Lee, *TRANSMIT: Summary of Incident Detection System Evaluation,* Washington, D.C., 1996.

Selected Bibliography

Ahmed, S. A., and A. R. Cook, "Time-Series for Freeway Incident Detection," *ASCE J. Transportation Engineering,* Vol. 106, No. 6, 1980, pp. 731–745.

Drew, D.R., Traffic Flow Theory and Control, New York: McGraw-Hill, 1968.

Tarnoff, P.J., and T. Batz, "TRANSCOM TRANSMIT project: A Unique Institutional Approach to a Unique Project," document obtained from TRANSCOM, 1996.

5
Incident Duration and Delay Prediction

This chapter comprises two major sections: incident duration prediction and delay prediction. Duration prediction is one of the most important steps of the overall incident management process. The real-time decisions regarding the resources needed to clear and manage the incident and the information given to travelers all depend on the knowledge of incident duration. An accurate and reliable estimate of the incident duration can be the main difference between an effective incident management operation and an unacceptable one. The delay estimation is also a crucial component of that real-time process. The decision regarding the initiation of traffic diversion especially depends heavily on the real-time delay estimation. Without knowing the delay that will be caused by an incident, it is impossible to decide whether diversion is warranted. However, for both duration and delay estimation, there is one major constraint: The duration and delay estimates should be obtained in real time. Just what does that constraint mean?

First, the parameters used in duration estimation models should be easy to obtain in real time. Some models incorporate variables, such as the age or income of drivers, that cannot be realistically obtained in real time under incident conditions. Thus, realistic incident duration estimation models need to be developed by keeping the real-time considerations in mind.

Second, the models should be fast enough to be run in real time. Real time in this case means a time period that is less than 10 to 15 minutes. That is especially important in the case of delay estimation. It is well known that popular deterministic queuing-based models are not as accurate as simulation-based stochastic models. However, simulation-based models are computationally expensive and almost all available simulation models require more than 10 minutes for multiple runs on the PC platform. It is important to note that when a simulation-based approach is used multiple runs are required to accurately capture the stochastic fluctuations. Thus, in a real-time decision-making process, it might not be always possible to use simulation as a realistic tool for incident delay estimation.

Third, as in the duration prediction, the number of parameters required for delay estimation is important for real-time usage of any delay estimation model. Simulation-based models require a large number of site-specific parameters that are not always easy to obtain for a large traffic network. On the other hand, the lack of accurate site-specific parameters reduces the reliability of a simulation model to a degree that it will not produce any useful estimates. It is important to understand that the simulation-based models are only as good as the parameters used as input. Time and resources needed to calibrate the parameters of a simulation model of a large transportation network can easily eliminate simulation from the list of feasible alternatives for real-time delay estimation alternatives.

This chapter is based on the research reported in [1–3]. It attempts to present the findings of that research conducted to develop practical yet reliable incident duration estimation models.

Section 5.1 describes various incident duration prediction models that have been developed by different researchers. Section 5.2 describes in detail the duration estimation study conducted in northern Virginia as part of the Wide-Area Expert GIS System FHWA Demonstration research project [1, 3]. The duration estimation models that have been developed as a result of that study have also been presented and compared with previous similar studies.

5.1 Incident Duration Estimation Models

This section reviews several major studies on incident data analysis and duration to summarize work in the area of incident duration estimation. Methodologies used in each study along with their findings and conclusions are also described.

In the study by Jones et al.[4] in the Seattle area, the primary source of accident duration data is the state police dispatch records over a 2-year period. The important concept in that study is one of conditional probability, in which the authors seek to determine the conditional probability that an incident lasts t minutes given that it has already lasted $(T-t)$ minutes where $t<T$. The incident duration is approximated by a log-logistic function as opposed to the log-normal distributions chosen elsewhere, even though the two are closely related.

The major drawback of the model is the use of several unrealistic variables in the model. For example, data such as the age of drivers are not easily obtained during real-time incident conditions. It is highly improbable and unpractical to assume the availability of such variables in predicting the clearance times when the incident clearance process is ongoing. The basis for the log-logistic function is specific to the Seattle region. The incident duration function is modeled as log-normal in some other cases. Golob et al.[5], in a specific study that deals with truck accidents, developed that argument further from theoretical considerations of the incident occurrence process. The analyses are organized according to four categories: type of collision, incident severity, incident duration, and lane closures.

Data from 9,508 truck accidents obtained from the Traffic Accident Surveillance and Analysis System (TASAS) in California was analyzed for the model development. In that data set, there are 4,436 personal injuries and 120 recorded fatalities. Relationships between the type of collision and selected truck accident characteristics are explored using the method of log-linear modeling. The first analysis is the position (highway or ramp) of occurrence of the various collision types. Overturns, broadsides, and hit objects are more common on ramps, while rear-end collisions and sideswipes are predominantly located on highways. The duration of an incident is hypothesized as comprising many sequential stages such as detection, initial response, injury attention, emergency vehicle response, accident investigation, debris, cleanup, and recovery.

It is assumed that the amount of time required for completion of each of those activities is influenced by the time required for the preceding activities. The duration of the nth activity is of the following form [5]:

$$Y_n - Y_{n-1} = Z_n Y_{n-1}, Z_n > 0 \tag{5.1}$$

where

Y_n = time of completion of nth response activity, measured from the start of the incident;

Z_n = random factor that relates duration of the nth activity to the cumulative time for the preceding activities.

Then

$$Y_n = Y_{n-1}(1 + Z_n) = Y_{n-2}(1 + Z_{n-1})(1 + Z_n) \tag{5.2}$$

$$Y_n = Y_0 \prod_{i=1}^{n} (1 + Z_i)$$

or

$$Y_n = Y_0 \prod_{i=1}^{n} W_i$$

where $W_i = (1 + Z_i)$, W_i 0.

By taking the logarithm of both sides, we obtain the following:

$$\ln Y_n = \ln Y_0 + \ln W_1 + W_2 + \ldots$$

That represents that the total incident duration is log-normally distributed. In the same study, for experimental purposes, a comparison is made between the log-normal and the log-uniform distributions. The log-normal distributions are shown to have higher relevance than the log-uniform distributions in terms of representing the incident duration distribution.

Another research effort for modeling freeway incident clearance time is part of a project that has been carried out at Northwestern University. That research [6] is an effort to develop an initial capability to provide successively improved incident clearance time predictions as time progresses. In that study, 121 incident records provided by the Illinois DOT Communications Center were used. Because the data set is not large enough, the *1988 Chicago Area Expressway Accidents Annual Report* (ACC), published by the Chicago Police Department, is used to examine the validity of the study data before applying the data to make any interpretation. A series of χ^2 tests shows that the study data set has a good representation and similar distribution to the large ACC data set. Incident reports are the original data used in the study. The first objective of the study is to find out what variables are important for the prediction of incident clearance time. Before making a detailed analysis for each variable, an overall incident clearance time regression model involving 22 variables is developed. Based on statistical assessment, 9 variables are found to show a strong relationship with the dependent variable, incident clearance time. Those statistically significant variables are determined as follows [6]:

- Operational factors: heavy wrecker (WRECKER), assistance from other response agencies (OTHER), sand/salt, pavement operations (SAND);
- Incident-type factors: number of heavy vehicles involved (NTRUCK), heavy loading (HEAVY), liquid or uncovered broken loadings in heavy vehicles (NONCON), severe injuries in vehicles (SEVINJ), freeway facility damage caused by incident (RDSIDE);

- Environmental factors: extreme weather conditions (WX).

Other variables are found to be statistically insignificant. Among them are number of vehicles, rollover vehicle, ratio of lanes blocked, number of severe injuries, day of week, distance from the central buisiness district (CBD), time of day, response time (RESP), second patrol arrival time, and incident report (HAR).

All the variables that are found to be statistically significant are included in the revised prediction model. The variables RESP and HAR, although statistically insignificant, are assumed to be important in improving incident operation study and thus are kept in the main model. The main model (with HAR) developed by the study is [6]

$$\text{CLEAR} = 14.03 + 35.57 \cdot \text{HEAVY} + 16.47 \cdot \text{WX} + 18.84 \cdot \text{SAND} - 2.31 \cdot \text{HAR} + 0.69 \cdot \text{RESP} + 27.97 \cdot \text{OTHER} + 35.81 \cdot \text{RDSIDE} + 18.44 \cdot \text{NTRUCK} + 32.76 \cdot \text{NONCON} + 22.90 \cdot \text{SEVINJ} + 8.34 \cdot \text{WRECKER (minutes)}$$

Furthermore, the concept of time-sequential incident clearance time prediction models is adopted in the study by Khattak et al. [7] because as time proceeds, further incident information becomes available for use in models, and more precise prediction results can be provided. A time sequence of the entire incident operation procedure and of those significant factors is defined. The variables are then grouped into four sequential incident clearance time prediction models. Predictions made by each model are based on the most important information that would be available before a specified time point. According to the availability of the information at different time stages, four sequential incident prediction models are developed.

However, the number of incidents considered for the analysis (109 in total) is not significant enough for the model results to be considered reliable. The validation of those results has also been done with only 12 incidents, which are too few for a reliable validation process.

Another recent study on incident clearance prediction is also conducted in Northwestern University as part of the ADVANCE Project task [8]. Information from 801 roadway incidents from Northwest Central Dispatch (NWCD) was used in the study. All the incidents are classified into the following incident types:

- Traffic stop/arrest (TSA);

- Motorist assist (stalled vehicles, flat tires, etc.) (MA);
- Accident with personal injuries (ACPI);
- Accidents involving property damage (ACPD);
- Severe incident (SI), including entrapment, load spills, and all fire-related incidents.

Preliminary data analysis is performed to find the distribution of incident clearance times for all incidents. It is found that most incident clearance times (more than half the sample) are very short, under 10 minutes, but a substantial fraction of incidents (19.8%) last more than 40 minutes, with some lasting more than an hour (10%). The preliminary analysis also examines the difference in average incident duration for different incident types, different roadway types, and different levels of congestion on the roadway, represented by time of day. The results of the data analysis show that incident duration is strongly associated with incident type, but differences in clearance times by time of day or roadway type are relatively small for each incident type.

Detailed analysis is used to determine the differences in incident types within each group, the roadway types (arterial vs. collector and local), the traffic conditions represented by the time of the day, and the intensity of the incident measured by the number of vehicles dispatched. Analysis of variance (ANOVA) is used to measure and test the statistical significance of differences in clearance times for each of these explanatory variables. Based on the statistical analysis results, an incident duration estimation decision tree is developed, as shown in Figure 5.1. In the figure, for each type of incident, an average incident duration time is determined based on the number of police cars dispatched (NP) and number of fire truck or ambulances dispatched (NF). For example, for ACPI incidents, three different average incident durations are given:

- For a total of NP and NF less than 3, the average incident duration is estimated as 57 minutes.
- If the total number of NP and NF is between 4 and 5, the average incident duration is estimated to be 61 minutes.
- Finally, if there are more than 6 NP and NF, the average incident duration is estimated as 74 minutes.

The approach used in the ADVANCE Project is simple, practical, and effective. However, the classification of incidents is oversimplified. For example, classification of severe incidents is not clear. Severe incidents may include

Incident Duration and Delay Prediction 89

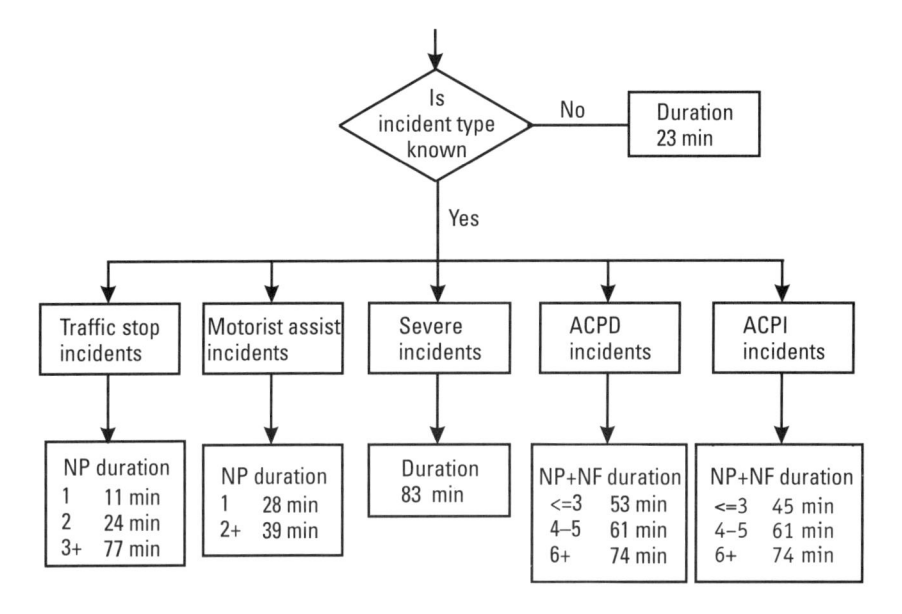

Figure 5.1 ADVANCE Project's decision tree for incident duration prediction. [*From:* [8] with permission from the authors.]

HAZMAT incidents, vehicle fires, truck incidents, and fatality incidents, and their corresponding clearance times may vary significantly. For each incident type, significant factors that affect clearance times are not fully explored. To better understand clearance characteristics, further study is needed.

Another research project sponsored by the FWHA has recently attempted to develop a model to predict freeway incidents and delays. The new model, called IMPACT, consists of four submodules [9]:

- The incident rate submodel estimates the annual number of incidents by type.
- The incident severity submodel estimates the number of lanes closed due to a specific incident and corresponding delays.
- The incident duration submodel estimates overall clearance time of an incident.
- The incident delay submodel estimates the delay caused by the incidents based on the prevailing traffic demand levels.

The duration submodel is developed using the log-normal distribution reported in the studies mentioned previously. The incident duration submodel is calibrated using data sets from Orlando and San Francisco I-880 [10], because they contained total incident durations, including detection and response times in addition to the incident clearance times. Mean and the standard deviation of the each log-normal distribution for different types of incidents are then calculated using those data sets.

Finally, [11] proposed an incident duration prediction model using the same I-880 data set used by [10]. In that data set, the exact start and end times of an incident are not known. Thus, the observed incident duration is corrected using an algorithm that takes the missing information into account [12]. In that data set, the following information for each incident is recorded:

- Incident type (accident, breakdown, etc.);
- Incident severity (number of lanes closed or affected);
- Vehicles involved (type, color, etc.);
- Time incident first witnessed;
- Response times (police arrival time, tow truck arrival time, clearance time, etc.);
- Weather conditions.

The second data set used in the study is related to the traffic characteristics along the study corridor. The data set consists of 30-second speed, flow, and occupancy data collected by the loop detectors installed along the freeway mainline and on/off ramp stations. A regression-based method is used to develop the duration estimation model. The final regression model has the following form [11]:

$$\text{Log}(duration) = 0.87 + 0.027 X_1 X_2 + 0.2 X_5 - 0.17 X_6 + 0.68 X_7 - 0.24 X_8 \quad (5.3)$$

where
  $duration$ = incident duration in minutes;
  X_1 = number of lanes affected by the incident;
  X_2 = number of vehicles involved in the incident;
  X_5 = dummy variable representing truck involvement in the incident (0 if truck is not involved; 1 otherwise);
  X_6 = dummy variable representing time of day (0 if morning peak time, 6:30 a.m.–9:30 a.m.; 1 if afternoon peak time, 3:30 p.m.–6:30 p.m.);

X_7 = natural logarithm of police response time (time from incident occurrence until police arrive at scene);

X_8 = dummy variable representing weather conditions (0 if no rain; 1 otherwise).

The two variables not used in this model, X_3 and X_4, are used in the incident delay prediction model presented in the same paper by Garib et al. [11]. The model is shown to be capable of predicting 81% of incident duration during morning and afternoon peak hours in a logarithmic format as a function of the six independent variables described in the preceding list. Moreover, it is shown that the addition of other variables does not significantly improve the accuracy of the predictions.

5.2 Northern Virginia Case Study: Methodological Structure

This section describes a comprehensive study conducted to determine the duration of incidents in northern Virginia. The purpose of this section is to give a complete review of a comprehensive study conducted to determine incident durations. We hope this comprehensive, step-by-step summary of a specific study helps the reader understand the details of the data collection and analysis efforts involved in such studies. This discussion is based on the information presented in [1, 3].

In the northern Virginia study, survey forms are used for investigating factors that influence incident clearance duration and for developing a model to predict incident clearance duration for different incident types. The methodology is summarized in Figure 5.2. There are several advantages to using surveys in the study. First, it is the only way to obtain the type of data needed. Incident management teams in northern Virginia do not collect detailed data for each incident. As a matter of fact, the only type of data that is collected in northern Virginia is the location, type, approximate starting and ending time of an incident, and the involvement of a tractor trailer. However, in this research, with the help of the domain expert, we have identified many parameters that might affect incident clearance times. Thus, survey research provides adequate richness of data and the ability to study factors that affect clearance times at a detailed level.

A shortcoming of using survey forms is the errors caused by the participants, namely, incident management personnel. Another important problem is the existence of inconsistencies among similar incident cases, which might be due to the operational differences among different agencies and individual teams in dealing with similar incident cases.

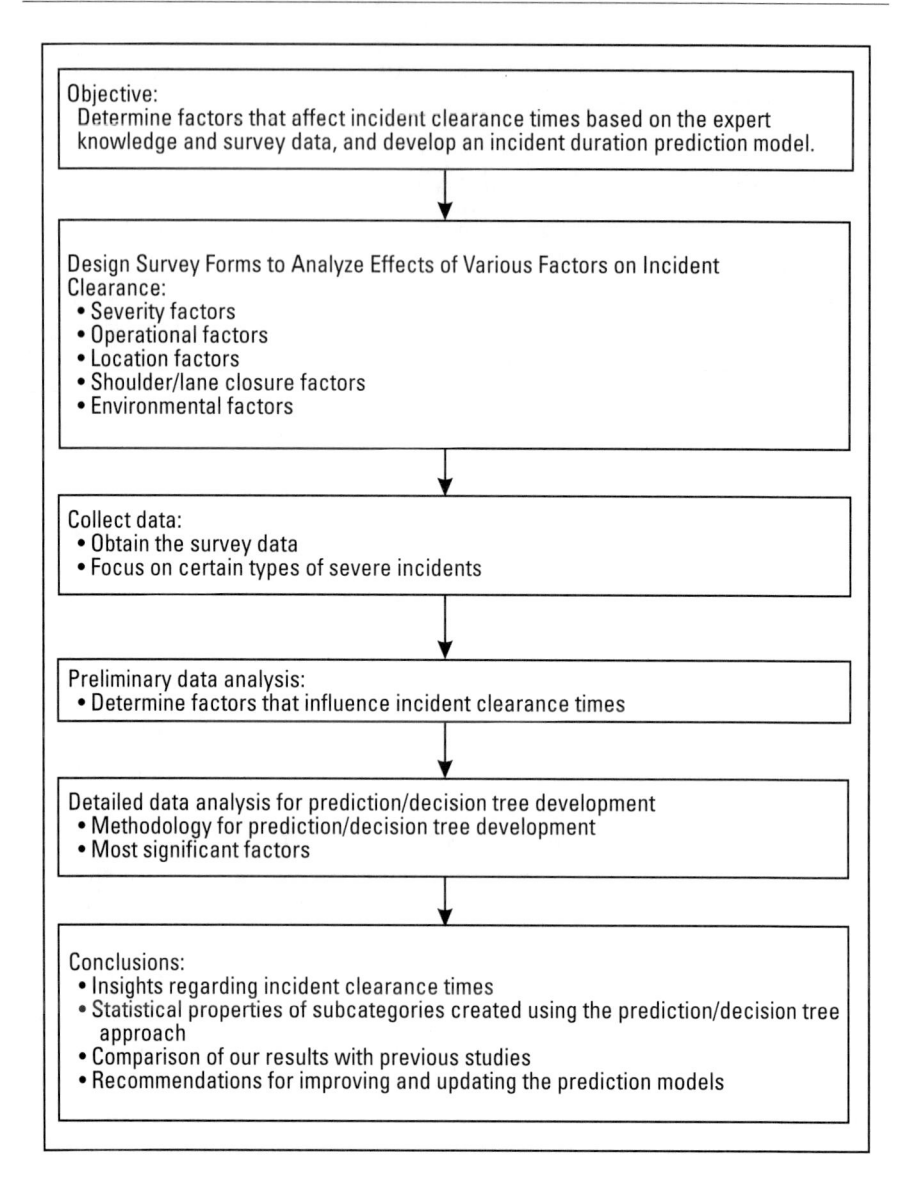

```
┌─────────────────────────────────────────────────────────────────┐
│ Objective:                                                       │
│ Determine factors that affect incident clearance times based on  │
│ the expert knowledge and survey data, and develop an incident    │
│ duration prediction model.                                       │
└─────────────────────────────────────────────────────────────────┘
                                  │
                                  ▼
┌─────────────────────────────────────────────────────────────────┐
│ Design Survey Forms to Analyze Effects of Various Factors on     │
│ Incident Clearance:                                              │
│   • Severity factors                                             │
│   • Operational factors                                          │
│   • Location factors                                             │
│   • Shoulder/lane closure factors                                │
│   • Environmental factors                                        │
└─────────────────────────────────────────────────────────────────┘
                                  │
                                  ▼
┌─────────────────────────────────────────────────────────────────┐
│ Collect data:                                                    │
│   • Obtain the survey data                                       │
│   • Focus on certain types of severe incidents                   │
└─────────────────────────────────────────────────────────────────┘
                                  │
                                  ▼
┌─────────────────────────────────────────────────────────────────┐
│ Preliminary data analysis:                                       │
│   • Determine factors that influence incident clearance times    │
└─────────────────────────────────────────────────────────────────┘
                                  │
                                  ▼
┌─────────────────────────────────────────────────────────────────┐
│ Detailed data analysis for prediction/decision tree development  │
│   • Methodology for prediction/decision tree development         │
│   • Most significant factors                                     │
└─────────────────────────────────────────────────────────────────┘
                                  │
                                  ▼
┌─────────────────────────────────────────────────────────────────┐
│ Conclusions:                                                     │
│   • Insights regarding incident clearance times                  │
│   • Statistical properties of subcategories created using the    │
│     prediction/decision tree approach                            │
│   • Comparison of our results with previous studies              │
│   • Recommendations for improving and updating the prediction    │
│     models                                                       │
└─────────────────────────────────────────────────────────────────┘
```

Figure 5.2 Methodology for incident duration prediction [3].

5.2.1 Structure and Design of Survey Forms and Data Collection

Collection and analysis of incident data were undertaken as part of the initial effort of the incident management project to understand incident clearance characteristics. The details of those efforts can be found in [1]. The aim of these subsections is to briefly summarize the early work for designing the survey forms and collecting the incident data. However, the emphasis in this chapter is on the analysis of new data and the development of incident clearance prediction/decision trees using the results of that analysis.

5.2.1.1 Design of Survey Forms and First Phase of Data Collection

Survey forms were prepared as a result of extensive meetings with the northern Virginia incident management experts and review of available incident management literature. Data were collected from incident clearance survey forms filled out by members of the participating agencies. The following agencies were involved in the project: Virginia State Police, Virginia DOT (VDOT) Safety Service Patrol, Fairfax County Police Department, and Fairfax County Fire and Rescue Department. About 10,000 survey forms were distributed to those agencies to be completed. Members were instructed on how to complete the survey forms, and over 1,000 personnel belonging to different agencies participated in completing the survey forms during roadway incidents. The data collection began in April 1994. Persons filling out the survey forms were asked to submit completed forms to their respective supervisors, who mailed the forms back to Virginia Tech. The information contained in the forms was then encoded into computer data files to be analyzed.

5.2.1.2 Samples of Data Collected

The information collected from the survey forms includes the name and agency of the person filling out the survey form; date and time of incident occurrence; and incident details such as the type and nature, number and type of vehicles involved, number of injuries and fatalities, prevailing weather conditions, location characteristics, lane closure information, resources used for clearing the incident, and the clearance time. A sample survey form is shown in Figure 5.3.

5.2.1.3 Incident Clearance Time

Incident clearance time is defined as the time from incident identification to the time the last emergency vehicle leaves the scene. Incident clearance time is used as a proxy for incident duration and should include the time from the beginning of an incident until the normal roadway capacity is restored. The time from the moment an incident occurs to the time the incident is verified and response starts is normally short in the northern Virginia area and does not vary significantly for different incident cases; thus, that time interval is ignored in our analysis. The time interval between the moment that an incident is cleared (response units leave the scene) to the time normal capacity is restored is affected by the traffic flow condition at that time and the number of delayed vehicles.

FREEWAY INCIDENT CLEARANCE SURVEY
Please see on the back side of this page for Instructions

Please fill in the Following about the Incident Response

Date: _____ Your Last Name: _____
Location: _____
Agency Affiliated to: _____
Event or Case # _____ Occurrence time of the Incident: _____
(OPTIONAL)

What was the Time of Occurrence of the Incident? Please Select one.

- 6:00 AM - 10:00 AM
- 10:00 AM - 12:00 NOON
- 12:00 NOON - 2:00 PM
- 2:00 PM - 4:00 PM
- 4:00 PM - 7:00 PM
- 7:00 PM - 1:00 AM
- 1:00 AM - 6:00 AM

What is the Incident Type? Select One. Choose the type based on the Major Causative.

Disabled Vehicle Property Damage Fatal Incident
Road Hazard Personal Injury Others

Is an involved Vehicle on Fire ? Yes No
Is there a HAZMAT involved? Yes No

If Yes, select the Type & Nature?

Hazmat State	Hazmat Nature	Hazmat Type
Solid	Spilled Fuel	Poisonous
Liquid	Spilled cargo	Radioactive
Gas	Engine Fluid Spill	Inflammable
	Cargo on Fire	Others
	Others	

Which one of the following Choices best Describes the Prevailing Weather?

Clear Misty Sleet/Ice
Cloudy Rain Smoke/Dust
Foggy Snow Other

Light & Temperature Select one each.

Bright < 45 deg
Satisfactory 45< t < 85
Dark > 85 deg

Please Select the appropriate choice for the following aspects of the Location

Land Use Type
Open Land Urban/Bldgs. Bridge/Tunnel

Please fill in the Following about the Incident.

# of Vehicles Involved _____
# of Cars Involved _____
# of Tractor Trailers _____
# of Personal Injuries _____
# of Fatalities Involved _____

Location Geometry (Select ALL appropriate Choices)

Straight Up Hill Elevated Rwy.
Curve Down Hill Ramps
Level

Resources Used for Clearance of Incident

Sweepers→ ___ # Front End Loaders→ ___
Spreaders→ ___ # Fire Engines→ ___
Cones→ ___ # Ambulances→ ___
Wreckers→ ___ # Arrow Boards→ ___
Sign Boards→ ___ Others (explain on back)→ ___
of PERSONNEL USED→ ___

HAZMAT Crew. Select One. Private Public None Used
Fixed Post Traffic Control Yes No
Alternate Route Established Yes No

What is your Opinion on the Resource Availability for Clearing this incident?

More than Adequate Adequate
Less than Adequate Absolutely Inadequate

Right Shoulder Left Shoulder
Present Absent Present Absent

Lane Closure Information.
Select the Lanes Blocked by the Incident.
Edit Lane Geometry presented to suit the Location.

Left Shoulder | Lane 6 | Lane 5 | Lane 4 | Lane 3 | Lane 2 | Lane 1 | Right Shoulder

What was the CLEARANCE TIME for the Incident? Select the best Alternative.

Below 15 Mins. Between 15 and 30 Mins. Between 30 and 45 Mins. Between 45 and 60 Mins. Between 60 and 75 Mins.

If above 75 Mins., Indicate the exact time below.

___ Hours and, ___ Mins.

Please Return Completed forms to your Supervisor. Thank You for Your Cooperation.

Figure 5.3 Sample survey form [1, 3].

5.2.1.4 Data Issues

The project addressed various issues involving the transfer of information from the returned survey forms into computer data files, the quality of the data, and the problems faced in extracting quality data from the survey forms.

As the first step in analyzing the incident data, information contained in the completed survey forms was transferred to computer data files. One of the biggest problems was ensuring the quality of the data. Many survey forms were incomplete. For example, in the case of incidents involving disabled vehicles, survey forms did not indicate the type of vehicle. Lane closure information, one of the most important factors influencing incident clearance times, was not properly filled out in a large number of survey forms. In some survey forms, the prevailing weather, light, and temperature conditions were missing, while in some others, more than one box was checked for these conditions.

Records with such incomplete and conflicting information about incident clearance, which if included would have seriously affected the outcome of the data analysis, were removed from the data sets. Only about one-half of the returned survey forms completed by the VDOT Safety Service Patrol were of sufficient quality to be usable. About 70% of the survey forms filled out by the Fairfax County Police Department were usable; 88% of the forms filled out by the Fairfax County Fire and Rescue Department were usable; and 97% of the forms filled out by the Virginia State Police were usable.

5.2.1.5 Experimental Design and Statistical Analysis of the Old Data (First-Phase Data Collection)

Statistical analysis software (SAS) was used for analyzing the data [1]. Incidents were grouped into six different categories: disabled vehicles, road hazards, property damage, personal injuries, fatalities, and other incidents.

As part of the preliminary analysis of the data, the average clearance times were calculated to compare the clearance times by incident type and agency. Table 5.1 shows the results of that analysis. The average clearance times by all data sets (agencies) for each incident type are as follows:

- Disabled vehicles: 7.5 minutes;
- Property damage: 33.6 minutes;
- Road hazard: 19.7 minutes;
- Personal injury: 41.9 minutes;
- Fatalities: 65.6 minutes;
- Other incidents: 15.2 minutes.

It can be seen that clearance times vary considerably with incident type and are comparable between agencies. In the case of the Fairfax County Fire and Rescue Department, the clearance times are a little higher because fire and rescue units are called only for more serious circumstances, thus increasing the clearance times.

Table 5.1
Average Clearance Times in Minutes by Incident Type and Agency [3]

Incident Type	Fairfax Co. Police Dept.	Fairfax Co. Fire&Rescue	Safety Service Patrol	VA State Police	Surveys from Sept. 1992
Disabled Vehicle	21.5 (46)	35.4 (7)	13.5 (373)	17.9 (739)	19.7 (400)
Road Hazard	36.0 (10)	— (0)	12.5 (24)	21.8 (43)	17.9 (46)
Property Damage	30.7 (203)	29.3 (51)	24.0 (59)	39.6 (165)	36.6 (135)
Personal Injury	40.7 (85)	33.7 (75)	52.5 (23)	53.6 (54)	39.6 (63)
Fatalities	— (0)	67.5 (1)	67.5 (1)	105.0 (2)	45.0 (4)
Other Incidents	16.5 (10)	22.5 (2)	9.5 (2)	14.4 (128)	16.3 (68)

* Numbers in parentheses are the number of incident cases.

The Fairfax County Police and Fire and Rescue departments dealt mostly with incidents involving property damage or personal injuries, while the great majority of incidents that involved Safety Service Patrol and Virginia State Police were disabled vehicles. Fewer than 1% of all incidents involved fatalities. Once the fact that incident type considerably affects clearance time was established, regression analyses were undertaken for each incident type to model clearance times on various factors thought to affect clearance time from consultations with experts. Clearance time for each incident type was modeled based on the following factors:

- Lane closure;
- Number of cars involved;
- Number of trucks involved;
- Number of personal injuries;
- Number of fatalities;
- HAZMAT involvement;
- Fire involvement;

- Time frame of occurrence;
- Prevailing weather;
- Prevailing temperature;
- Land-use type.

That statistical analysis is conducted to study the effect of the various "hypothesized" variables on the clearance of incidents. As expected, several of those variables have significant effects on clearance times, which follow different priorities or patterns, depending on the incident type. In general, the study clearly establishes the fact that the clearance times of incidents are affected by the incident types. That was evident from the clearly different mean and median values of clearance times obtained. In addition, the low variances observed reinforces the representativeness of the mean in the data sets. As part of the study, linear regression models are developed to model the incident duration. These models, unfortunately, have very low R-square values; the possible reasons for that are explained in the same report. Based on the results of those surveys and expert advice, it was decided that the second phase of the surveys would commence on July 15, 1994 and would be different from the initial survey effort.

5.2.1.6 New Survey Form

The second phase of data collection started in July, 1994 and continued until the beginning of 1995. Survey forms were filled out, and data analysis aimed at developing models to predict incident clearance times was developed at Virginia Tech [1]. Based on the preliminary statistical data analysis and examination of problems found in the forms during data entry and based on feedback obtained from the agencies involved in the survey, a new survey form was designed to try to obtain better results. The following are some of the major changes implemented in the new survey:

- Removal of ambiguous and redundant questions;
- Addition of questions that are more relevant based on old survey results;
- Elimination of open choices and changing question options to force responses into the study variable categories or requirements;
- Clarification of instructions to minimize incomplete responses and improve responsiveness;

- Determination to discontinue survey of disabled cars on shoulders of roadways since ample data are already available and to reduce workload on respondents.

Various changes were made to the old survey form to generate the new one. The reasons for the changes and a copy of the new survey form are presented in [1]. The most important changes made on the new form are:

- Splitting the category of disabled vehicles into disabled car and disabled truck;
- Omitting information on disabled-car-on-shoulder incidents since normal disablements on shoulders account for over 60% of all incidents and plenty of data are already available;
- Adding disabled truck and disabled car in travel lane as two different incident types instead of disabled vehicle item on the old form;
- Adding a question regarding exact clearance time instead of time intervals motivated by the statistical analysis of the old data.

5.2.2 Analysis of New Incident Data

Over 900 new forms have been filled out by northern Virginia incident management personnel and returned to the Center for Transportation Research at Virginia Tech during the second phase of data collection. The data items with serious errors or with vital attributes (e.g., incident type) missing were eliminated during the data entry process. As a result of that processing, a new data set consisting of 650 incidents was obtained. The new data set was used for the data analysis and the development of decision/prediction trees described in this section.

First, linear regression technique is used to model the incident duration. However, due to the wide variation of clearance time data, a low R-square value (approximately 0.35) is obtained by using different combinations of variables in the prediction model. The data also are tested to determine if it follows either log-normal [5] or log-logistic distributions [4]. The results of statistical significance tests show that the data do not follow either of those distributions. Next, new approaches for the prediction, including neural networks and prediction/decision trees, are undertaken. Among those approaches, the decision/prediction tree approach has provided a good and simple way of modeling the duration prediction problem with the available data.

The following subsections describe the methodology adopted for developing the prediction/decision trees. Then the variables considered are described, followed by the preliminary and detailed data analysis results. Next, results and conclusions of the data analysis are presented and discussed. Finally, prediction/decision trees developed to estimate incident clearance times and their validation are presented.

5.2.2.1 Methodology for Developing Prediction/Decision Trees

The methodology adopted for developing prediction/decision trees is similar to the one used by the ADVANCE Project [8]. However, there are some major improvements [2]:

- The data set does not contain normal car disablements on the shoulder and minor traffic violation stop and arrest incidents. Such incidents account for the majority of the total incidents and do not have much impact on traffic. Consistent results have been achieved in the first stage of data analysis [1].
- An improved and more detailed incident type classification method is used in this study.
- A larger number of significant variables have been explored in each incident type group.

The methodology for developing prediction/decision trees is a simple one. The process first considers the complete data set X. It then examines all possible splits of X into two or more subsets, based on the values of the independent variables and the selection of best split. The decision for the best split at each step is made by using ANOVA, and a tree is then constructed through a series of splits. The process of constructing a prediction/decision tree is similar to the classification/regression tree approach (CART) described in detail in [13] and in [3]. The major difference is that CART uses a binary split, while our classification methodology allows nonbinary splits. Also, CART uses the mean squared error as the cost function associated with a tree having K terminal nodes. Each binary split in CART is chosen as the split that produces the greatest reduction in mean squared error cost. That is also similar to using ANOVA, which can be described as a technique to analyze total variation among different populations.

5.2.2.2 Description of Variables

The following are candidate variables considered to have an impact on incident clearance times.

Incident-Type Factors

    INCTYPE: Incident type
    INCTYPE = 1: Road hazard incident
    INCTYPE = 2: Property damage incident
    INCTYPE = 3: Personal injury incident
    INCTYPE = 4: Vehicle fire incident
    INCTYPE = 5: Fatal incident
    INCTYPE = 6: Cargo spill incident
    INCTYPE = 7: HAZMAT incident
    INCTYPE = 8: Weather-related incident
    INCTYPE = 9: Construction/maintenance
    INCTYPE = 10: Disabled car in travel lane incident

Incident Detail/Severity Factors

    NCARS: Number of cars involved in the incident
    NTRUCKS: Number of trucks involved in the incident
    NINJUR: Number of injuries involved in the incident
    NFATAL: Number of fatalities involved in the incident

Operational Factors

    POLICVEH: Number of police vehicles on the scene
    FIREENG: Number of fire engines on the scene
    AMBUL: Number of ambulances on the scene
    WRECKER: Number of wreckers used
    ARRBRDS: Number of arrowboards used
    MEDVAC: MEDVAC helicopter requested? (y/n)
    FIXPOST: Fixed post-traffic control? (y/n)
    ALTROUT: Alternative route established? (y/n)

Location Factors

    RWTYPE: Roadway type
    RWTYPE = 1: Freeway
    RWTYPE = 2: Nonfreeway
    LANDUSE: Land use type
    LANDUSE = 1: Open land

LANDUSE = 2: Urban/buildings
LANDUSE = 3: Bridge
LANDUSE = 4: Tunnel

Shoulder/Lane Closure Factors

SHOULDER: Shoulder information
SHOULDER = 1: Shoulder presence
SHOULDER = 0: Shoulder absence
LANCLOSR: Ratio of lanes closed to total number of lanes

Environmental Factors

WEATHER: Weather condition
WEATHER = 1: Clear/cloudy
WEATHER = 2: Rain
WEATHER = 3: Foggy/misty
WEATHER = 4: Snow
WEATHER = 5: Sleet/ice
LIGHT: Light condition
LIGHT = 1: Bright
LIGHT = 2: Satisfactory
LIGHT = 3: Dark
TEMP: Temperature
TEMP = 1: 45 degrees
TEMP = 2: 45–85 degrees
TEMP = 3: 85 degrees

The variable CLT is the value of incident clearance time in minutes.

5.2.2.3 Preliminary Analysis

The frequency distribution of incident clearance times and summary information for all incidents are shown in Figure 5.4. The figure shows that most incident clearance times are within 1 hour but that a substantial fraction (15.0%) of incidents last more than 1 hour and some severe incidents (about 0.8%) last more than 3 hours. The mean is 44 minutes, and the standard deviation is 33.8 minutes. The distribution is skewed as opposed to approximately normal in the case of some other studies, such as [4]; that may be because our data set does not contain a large number of longer incidents. Further, the data set at this

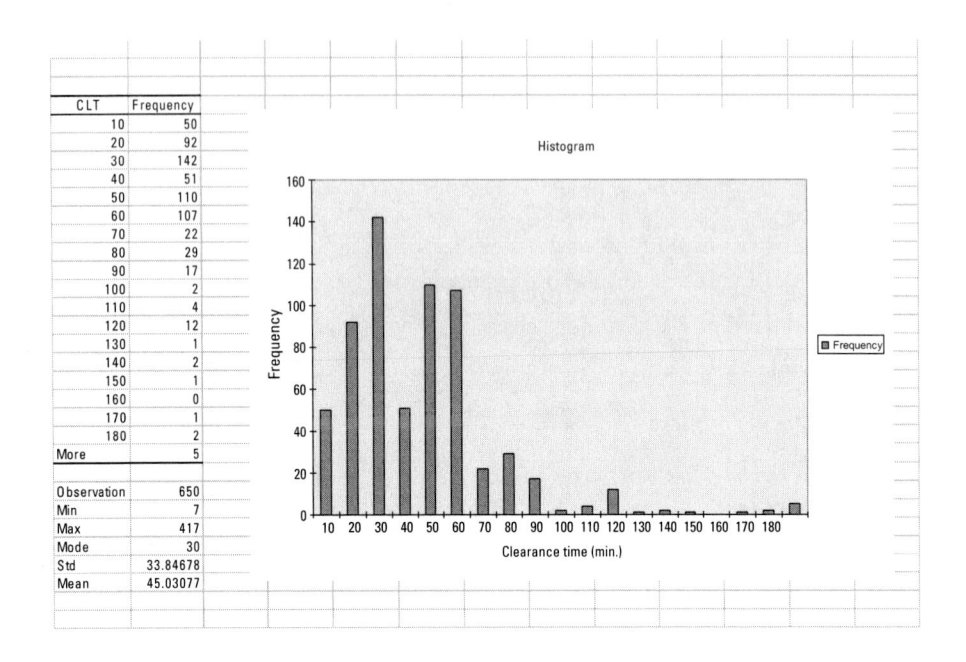

102 Incident Management in Intelligent Transportation Systems

CLT	Frequency
10	50
20	92
30	142
40	51
50	110
60	107
70	22
80	29
90	17
100	2
110	4
120	12
130	1
140	2
150	1
160	0
170	1
180	2
More	5
Observation	650
Min	7
Max	417
Mode	30
Std	33.84678
Mean	45.03077

Figure 5.4 Incident clearance time distribution [2, 3].

point is not homogenous enough and contains many types of incidents. Most of the previous incident duration studies where normal or approximately normal distributions are found concentrated on one type of incident such as truck incidents [12].

The first step of the preliminary analysis is to determine the characteristics of incidents that have different clearance times and to examine the differences in average incident clearance times for different incident types and different roadway types. The frequency distributions of incident clearance times for freeway and nonfreeway incidents are presented in Figures 5.5 and 5.6 respectively. The frequency distributions of incident clearance times for different types of incidents are presented in Figures 5.7–5.12. In the whole data set, there are only three cases of HAZMAT incidents, with clearance times of 135, 180, and 417 minutes. Only two cases of weather-related incidents are recorded, with clearance times of 15 and 150 minutes. Therefore, we do not have enough data points to analyze HAZMAT and weather-related incidents. Table 5.2 summarizes average clearance times by incident type.

The results of preliminary analysis show that incident clearance times are strongly associated with incident types, but differences in clearance times by roadway type are relatively small. Both the average values and the distribution

Incident Duration and Delay Prediction 103

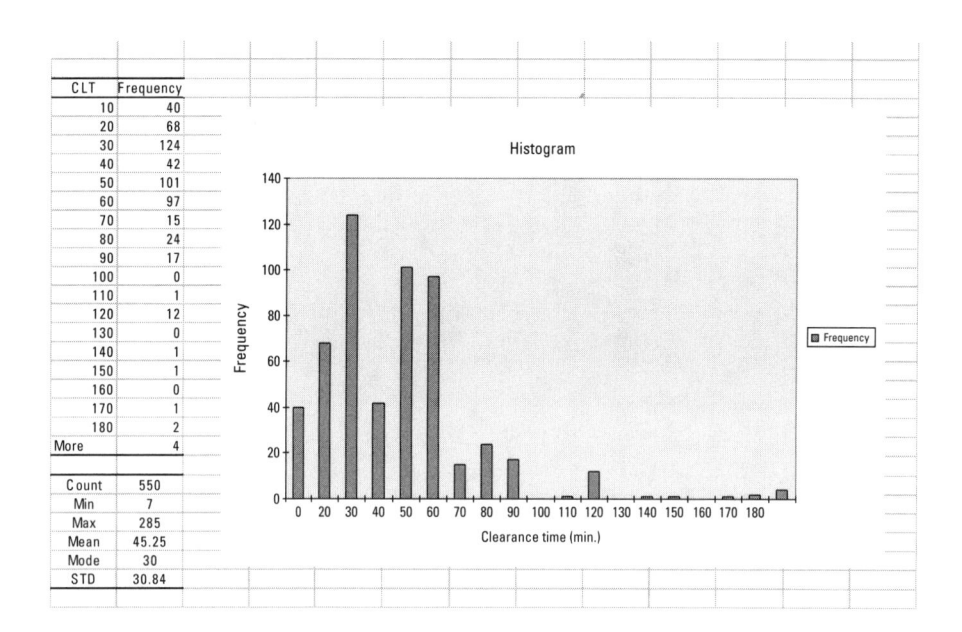

CLT	Frequency
10	40
20	68
30	124
40	42
50	101
60	97
70	15
80	24
90	17
100	0
110	1
120	12
130	0
140	1
150	1
160	0
170	1
180	2
More	4

Count	550
Min	7
Max	285
Mean	45.25
Mode	30
STD	30.84

Figure 5.5 Distribution of freeway incident clearance times [2, 3].

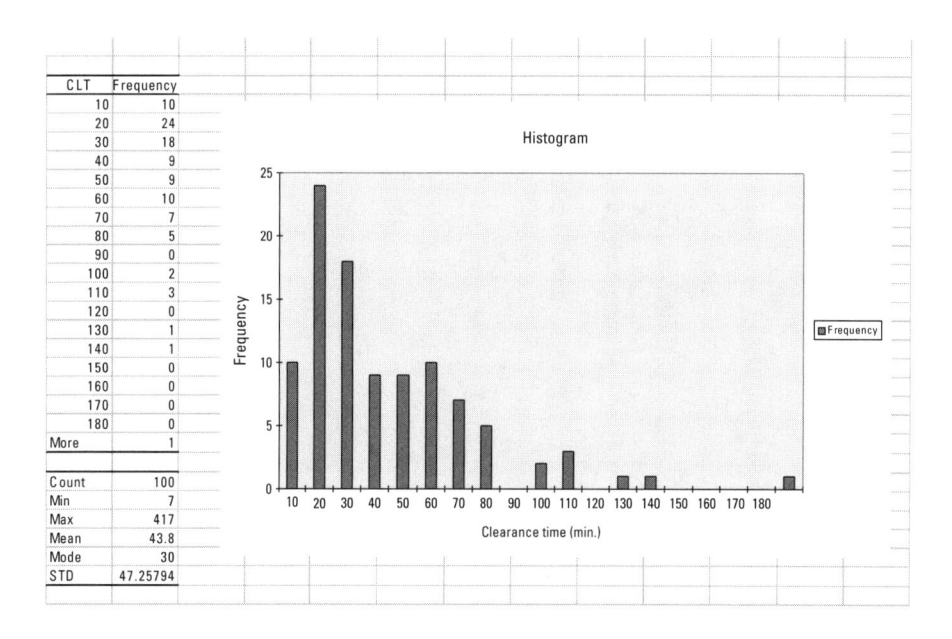

CLT	Frequency
10	10
20	24
30	18
40	9
50	9
60	10
70	7
80	5
90	0
100	2
110	3
120	0
130	1
140	1
150	0
160	0
170	0
180	0
More	1

Count	100
Min	7
Max	417
Mean	43.8
Mode	30
STD	47.25794

Figure 5.6 Distribution of nonfreeway incident clearance times [2, 3].

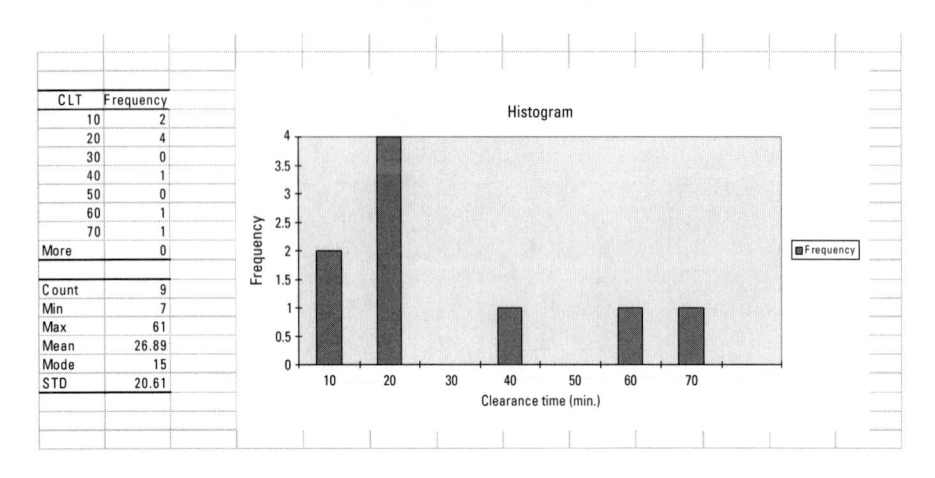

Figure 5.7 Distribution of road hazard incident clearance times [2, 3].

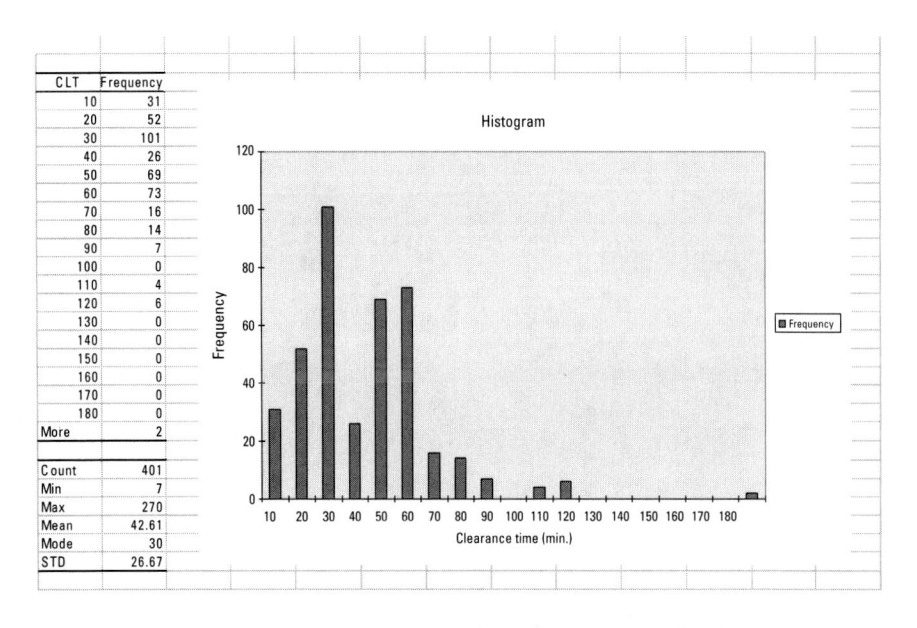

Figure 5.8 Distribution of property damage incident clearance times [2, 3].

Incident Duration and Delay Prediction

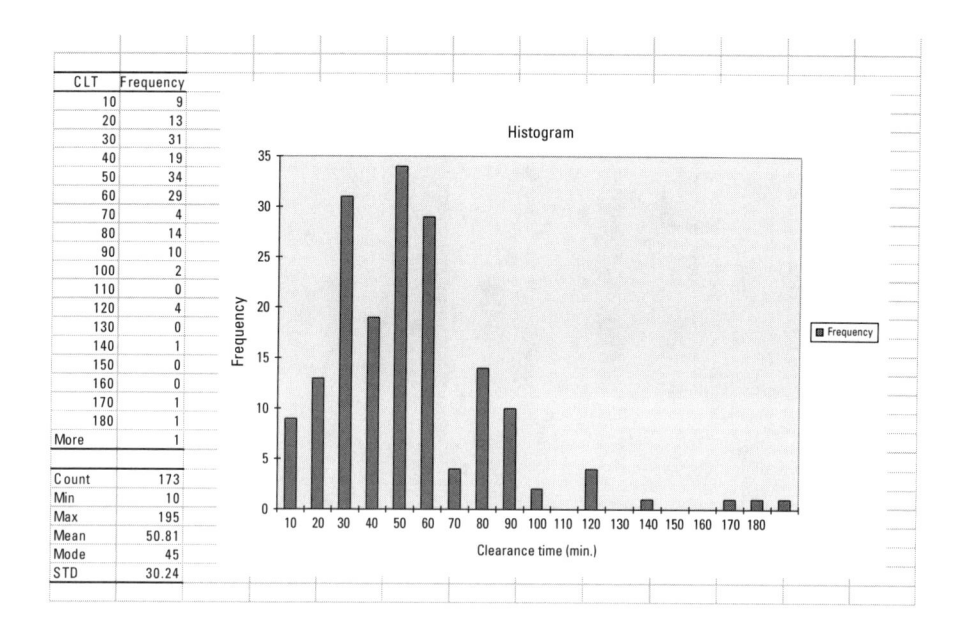

CLT	Frequency
10	9
20	13
30	31
40	19
50	34
60	29
70	4
80	14
90	10
100	2
110	0
120	4
130	0
140	1
150	0
160	0
170	1
180	1
More	1

Count	173
Min	10
Max	195
Mean	50.81
Mode	45
STD	30.24

Figure 5.9 Distribution of personal injury incident clearance times [2, 3].

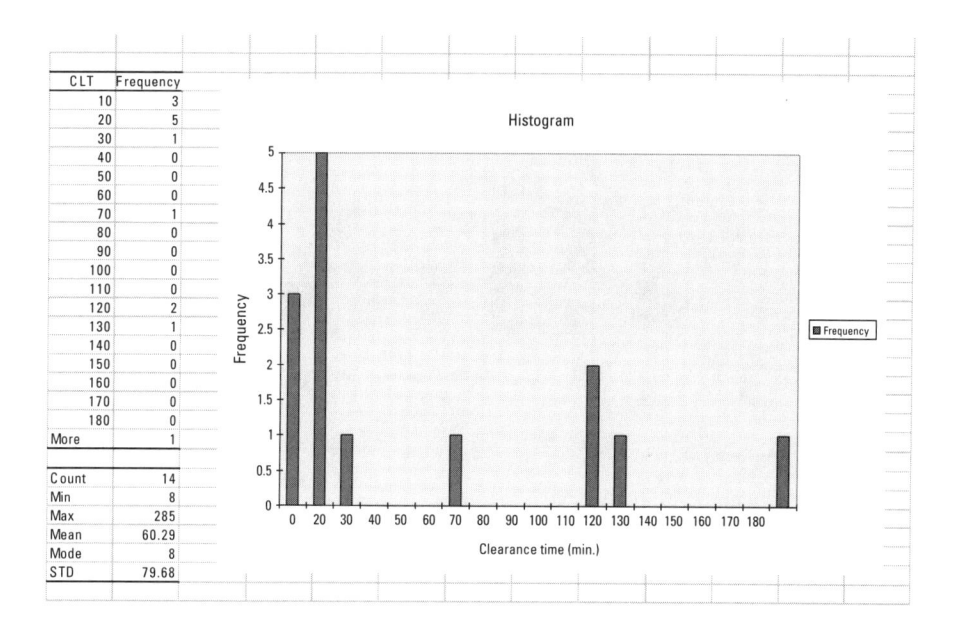

CLT	Frequency
10	3
20	5
30	1
40	0
50	0
60	0
70	1
80	0
90	0
100	0
110	0
120	2
130	1
140	0
150	0
160	0
170	0
180	0
More	1

Count	14
Min	8
Max	285
Mean	60.29
Mode	8
STD	79.68

Figure 5.10 Distribution of disabled truck incident clearance times [2, 3].

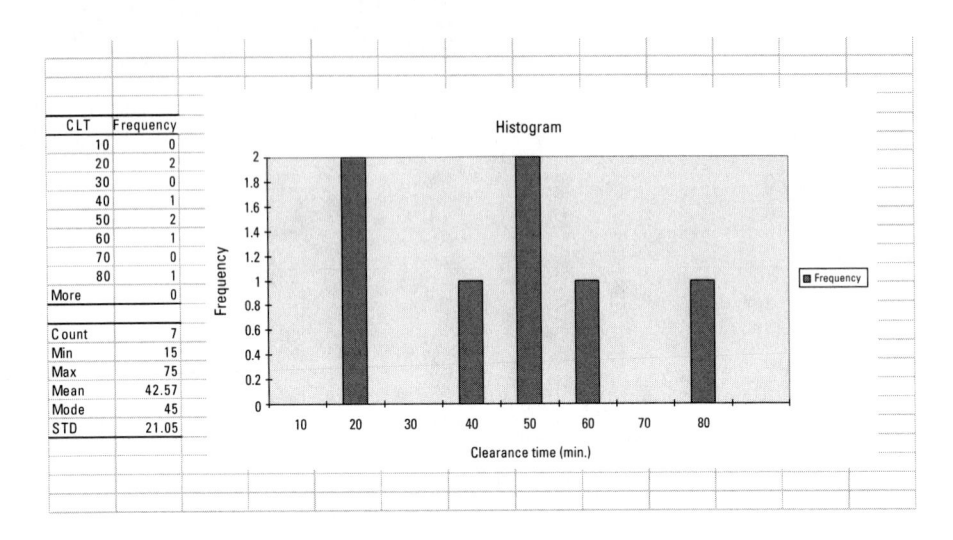

CLT	Frequency
10	0
20	2
30	0
40	1
50	2
60	1
70	0
80	1
More	0

Count	7
Min	15
Max	75
Mean	42.57
Mode	45
STD	21.05

Figure 5.11 Distribution of vehicle fire incident clearance times [2, 3].

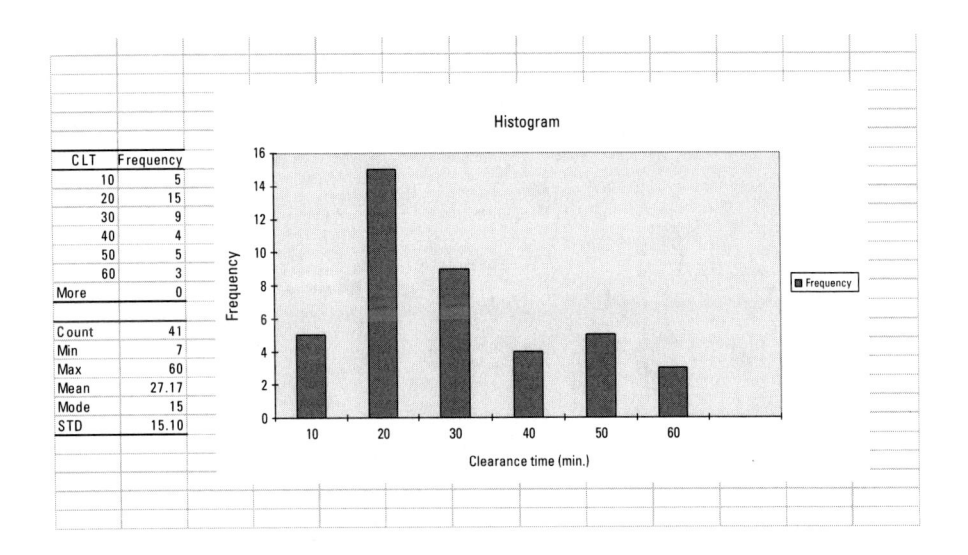

CLT	Frequency
10	5
20	15
30	9
40	4
50	5
60	3
More	0

Count	41
Min	7
Max	60
Mean	27.17
Mode	15
STD	15.10

Figure 5.12 Distribution of disabled car in travel lane incident clearance times [2, 3].

Table 5.2
Average Clearance Times by Incident Type [2, 3]

Incident Type	Road Hazard	Property Damage	Personal Injury	Disabled Truck	Vehicle Fire	HAZMAT	Weather Related	Disabled Car	All
Mean (min.)	26.89	42.61	50.81	60.29	42.57	244.00	82.50	27.17	45.03
# Cases	9	401	173	14	7	3	2	41	650

patterns differ significantly for each incident type for most cases. For example, the average clearance time for HAZMAT incidents is over 6 hours, while car disablements last, on average, less than one-half hour. The ANOVA results also strongly suggest the need to consider the incident type as the major factor for additional analysis.

Further, the results of the preliminary analysis of the incident clearance times indicate that there is a wide distribution of incident clearance times even for incidents of the same type. That supports the need for additional analysis using several factors, such as incident type, severity of incidents, operational factor, roadway type, and environmental factors.

5.2.3 Detailed Analysis

The aim of the detailed analysis is to determine the effects of factors such as incident type, incident severity, operational factor, location, lane/shoulder closure, and environmental factors on the incident clearance times. ANOVA is used to measure and test the statistical significance of differences in incident clearance times for each of those explanatory variables. Prediction/decision trees are then built based on the results of ANOVA analysis. For groups that do not have a large enough sample size to run statistical significance tests, average values and an interval of 25 and 75 percentiles as lower and upper limits, respectively, are used for the purpose of prediction/decision tree development.

5.2.3.1 Road Hazard

The first type of incident type that is analyzed is road hazard incidents. Because there are only nine cases of road hazard incidents, statistical significance tests cannot be carried out. Average values and 25 and 75 percentiles are used as estimates of the clearance times.

5.2.3.2 Property Damage

The second type of incident that is studied is property damage. Many incident detail/severity factors described in Section 5.1 that might affect the clearance duration of property damage incidents are considered. The impact of those factors on clearance time is investigated using ANOVA tests. Among those factors, truck involvement and number of vehicles involved are found to affect the incident clearance times the most. The results of ANOVA tests show that truck involvement has more impact on clearance times than the number of vehicles involved. Thus, the data is first grouped into two subcategories: incidents with truck involvement and incidents without truck involvement.

The next step is to determine other significant factors for each subcategory using ANOVA. It is found that the number of vehicles involved in the incident is a significant factor that affects clearance times of property damage incidents without truck involvement, especially when the number of vehicles involved is four or more, in which case the clearance time increases significantly. However, in the subgroup of property damage incidents with truck involvement, neither the number of trucks involved nor the number of vehicles involved turns out to be significant.

Operational Factors

Among the operational factors, number of police vehicles on scene (POLICEVEH) is found to be significant for both the subcategories of number of cars involved, namely, 1–3 cars involved and 4 or more cars involved. Table 5.3 shows the impact of number of police vehicles involved on clearance time estimations for property damage incidents without truck involvement.

Table 5.3
Impact of Number of Police Vehicles Involved on CLT Prediction [2, 3]

No. of Vehicles Involved	POLICEVEH	Extreme Low	Extreme High	Average Value	Lower Limit of the Estimate Interval	Upper Limit of the Estimate Interval
1-3	1	7	120	38	25	50
	2	10	90	46	35	60
	3, or 3+	10	120	65	45	90
4, or 4+	1	15	80	39	20	60
	2, or 2+	30	270	97	60	110

The next step is to study the effect of other agencies' involvement on property damage incident clearance times. The involvement of other agencies is reflected by the response of wreckers, ambulances, and fire engines.

For property damage incidents, only the impact of wrecker involvement is considered, since ambulances normally are requested for personal injury incidents, and fire engines are requested only for vehicle fire incidents. The results show that wrecker and ambulance involvement significantly increases incident clearance times for the cases of 1–3 vehicles involved and 1 police car involved. However, for the cases of more than 1 police vehicle involved or more than 3 vehicles involved, the impact of wrecker involvement is either statistically insignificant or unknown due to insufficient sample size.

The other operational factors found to be important while studying the whole data set are arrowboard used or alternative routes established. However, due to the limited data points at this subcategory, the statistical significance tests cannot be performed.

Location Factors

The two location factors that have an impact on the clearance time are found to be roadway type and land use type. Incidents that occur on freeways last a shorter time than arterial incidents, although the difference is very small and thus ignored. Incidents that occur in rural areas last a little longer than urban incidents, but again, the difference is very small. The only location factor that significantly affects clearance time is found to be LANDUSE3, that is, an incident on a bridge or in a tunnel. Incidents on bridges or in tunnels take, on average, about 15 minutes more to clear. In this data set, however, few incidents have that property. This result will be included in the final knowledge base as a heuristic rule that differentiates regular incidents from incidents occurring in tunnels or on bridges.

Environmental Factors

The only environmental factor that affects clearance times is an inclement weather condition. The incidents occurring under inclement weather conditions take, on average, about 7 minutes longer to clear.

5.2.3.3 Personal Injury Incidents

A similar methodology is followed to develop prediction/decision trees for estimating incident clearance durations for the personal injury incidents.

Incident Severity/Detail Factors

ANOVA tests show that both truck involvement and the number of vehicles involved are statistically insignificant. However, the impact of the number of injuries is found to be significant for personal injury incidents. Thus, the incident clearance time data are grouped into two subcategories: incidents with 1–2 injuries and incidents with 3 or more injuries.

Operational Factors

The number of police vehicles on the scene (POLICEVEH) is found to be significant for both the subcategories of 1–2 injuries and 3 or more injuries. Table 5.4 shows the impact of the number of police vehicles involved on clearance time prediction for personal injury incidents. The data analysis shows an increase in average incident clearance times as a function of the number of injuries and the number of police vehicles. That is an expected result because incident clearance times are expected to be higher for more severe incidents, where the number of injuries and responding police vehicles are higher than less severe incidents.

The next step is to study the other agencies' involvement on incident clearance times. The involvement of other agencies is reflected by the involvement of wreckers, ambulances, and fire engines. For personal injury incidents, only the impact of wrecker and ambulance involvement is considered, since fire engines normally are requested for vehicle fire incidents. The results show that the total number of wreckers and ambulances is a significant factor that affects incident clearance times for the cases of 1–2 injuries and 1 police car involved. However, for the cases of more than 1 police vehicle involved or more than 2 injuries, the impact of the total number of wreckers and ambulances is either statistically insignificant or unknown due to insufficient sample size. Table 5.5 shows the impact of the total number of wreckers and ambulances on clearance times for personal injury incidents with 1–2 injuries and 1 police car involved.

Table 5.4
Impact of the Number of Police Vehicles Involved on CLT Prediction for Personal Injury Incidents [2, 3]

No. of Injuries	POLICEVEH	Extreme Low	Extreme High	Average Value	Lower Limit of the Estimate Interval	Upper Limit of the Estimate Interval
1–2	1	10	120	42	31	60
	2	10	135	52	35	60
	3, or 3+	20	90	60	35	69
3, or 3+	1, 2	20	100	56	60	80
	3, or 3+	60	195	107	75	160

Table 5.5
Impact of Total Number of Wreckers and Ambulances on CLT for Personal Injury Incidents With 1–2 Injuries and 1 Police Car Involved [2, 3]

WRECKER + AMBULANCE	Extreme Low	Extreme High	Average Value	Lower Limit of the Estimate Interval	Upper Limit of the Estimate Interval
0–1	10	95	35	20	45
2, or 2+	10	120	51	35	60

Again, the finding is not surprising because the total number of wreckers and ambulances needed to clear the incident is definitely higher for more severe incidents.

Similar to the analysis of property damage incidents, some other operational, location, and environmental factors are found to be statistically significant for the whole personal injury data set. However, due to the limited data points at each subcategory, the statistical significance test cannot be performed at that level. Some adjustments to the mean values and interval limits were made according to the statistical analysis results on the whole personal injury data set. Those values are shown on the final decision trees.

5.2.3.4 Disabled Truck Incidents

Because there were only 14 cases of disabled truck incidents, carrying out a statistical significance test is unnecessary. Therefore, average values, 25 and 75 percentiles, are used as estimates of the clearance times. From the data analysis, it was found that the variance of clearance times is quite large for those 14 cases of disabled truck accidents. On the other hand, it was observed from data that wrecker involvement increased clearance time significantly, from 32 minutes for no wrecker involvement to 76 minutes for wrecker involvement. That finding once more showed that involvement of operational vehicles such as wreckers is a good indication of higher clearance times. However, due to lack of data, no further statistical significance analysis can be performed. Thus, it is not possible to statistically validate the results of this analysis until more data are obtained in other studies.

5.2.3.5 Vehicle Fire Incidents

Because there were only 7 cases of vehicle fire incidents in our database, carrying out a statistical significance test was not possible. Average values of 43 minutes, 25 and 75 percentiles, 38 and 40 minutes were used as estimates of the clearance times.

5.2.3.6 HAZMAT Incidents

Because there were only 3 cases of HAZMAT incidents, carrying out a statistical significance test is not possible in this case. An average of the 3 incidents, 244 minutes, was used as the default estimate.

It is clear that more data are needed for severe incidents like HAZMAT incidents. However, the occurrence of that type of incident is rare, so data are not readily available.

5.2.3.7 Weather-Related Incidents

There were only 2 cases of weather-related incidents, with clearance times of 15 minutes and 150 minutes. Thus, it is not possible to comment on this type of incident.

5.2.3.8 Disabled Car in Travel Lane

There were 41 cases of disabled car in travel lane incidents, but no statistically significant factor on clearance times can be identified. Average value of 27 minutes, 25 and 75 percentiles, 20 and 40 minutes were used as estimates of the clearance times.

5.2.4 Summary of Detailed Data Analysis

Property damage and personal injury incidents are the majority of the incident data collected and analyzed in this study. Due to the large sample size of the data of these two incident types, detailed significance analyses are performed for each category. For road hazard, vehicle fire, disabled truck, and disabled car in travel lane incidents, the sample sizes are small; however, small variations and consistent results are found for these incident types. Therefore, clearance time prediction can be made with acceptable accuracy based on the conclusions drawn from the fairly small data samples and the expert knowledge. For HAZMAT and weather-related incidents, the number of incidents for each category is too small to draw any meaningful conclusions. More data and studies are needed to determine the factors that affect the clearance times of these incident types.

5.2.5 Development of Incident Clearance Time Prediction/Decision Trees

Based on the detailed analysis described in Section 5.2.3, prediction/decision trees for incident clearance time prediction have been developed. The first split of the overall data set is made according to incident type, then all the other significant affecting factors are considered and the data are further split into smaller subsets. Figures 5.13 through 5.16 show the final decision trees. As mentioned before, some factors have been found to be important but are not included in the decision trees due to an insufficient number of cases. Adjustments to the clearance times can be made on factors such as the use of arrowboards, the establishment of alternative routes, the existence of inclement weather conditions, and the occurrence of incidents on a bridge and in a

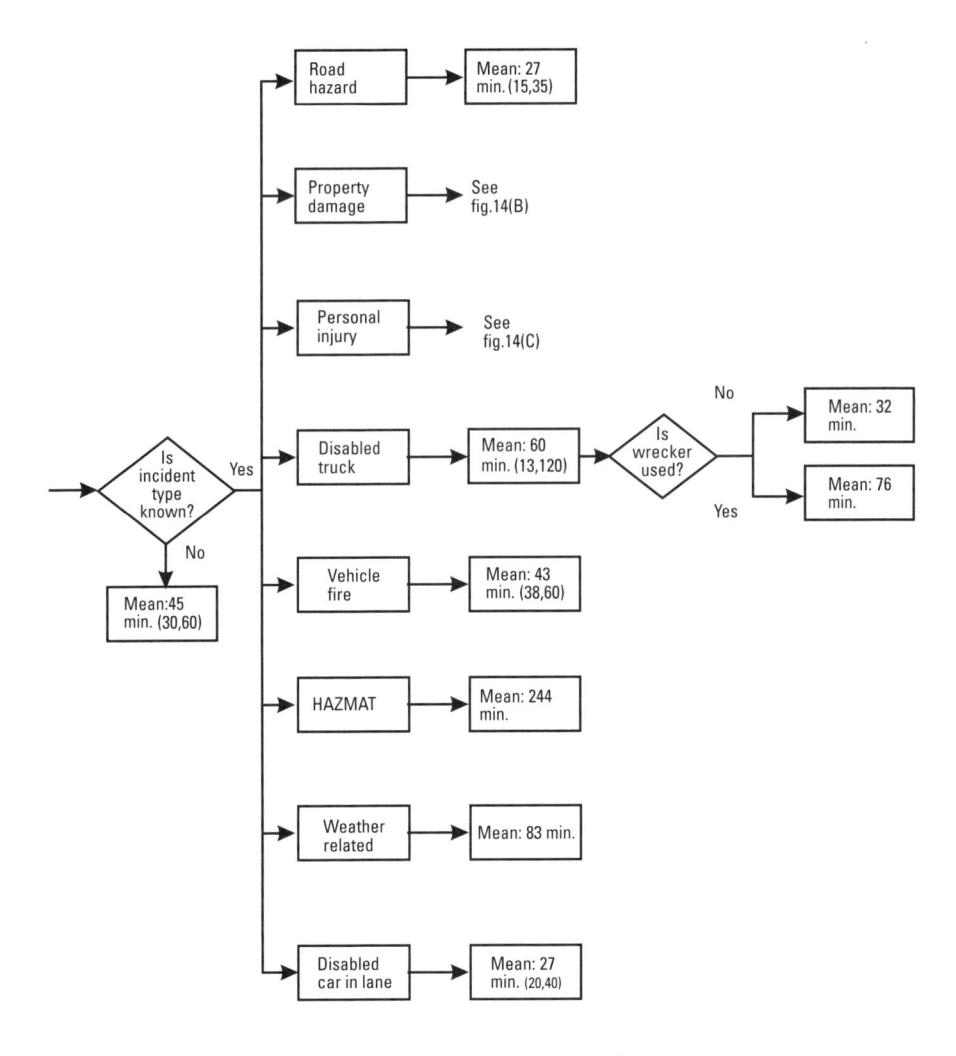

Figure 5.13 Decision tree for incident clearance time prediction (numbers in parenthesis are interval limits in minutes) [2, 3].

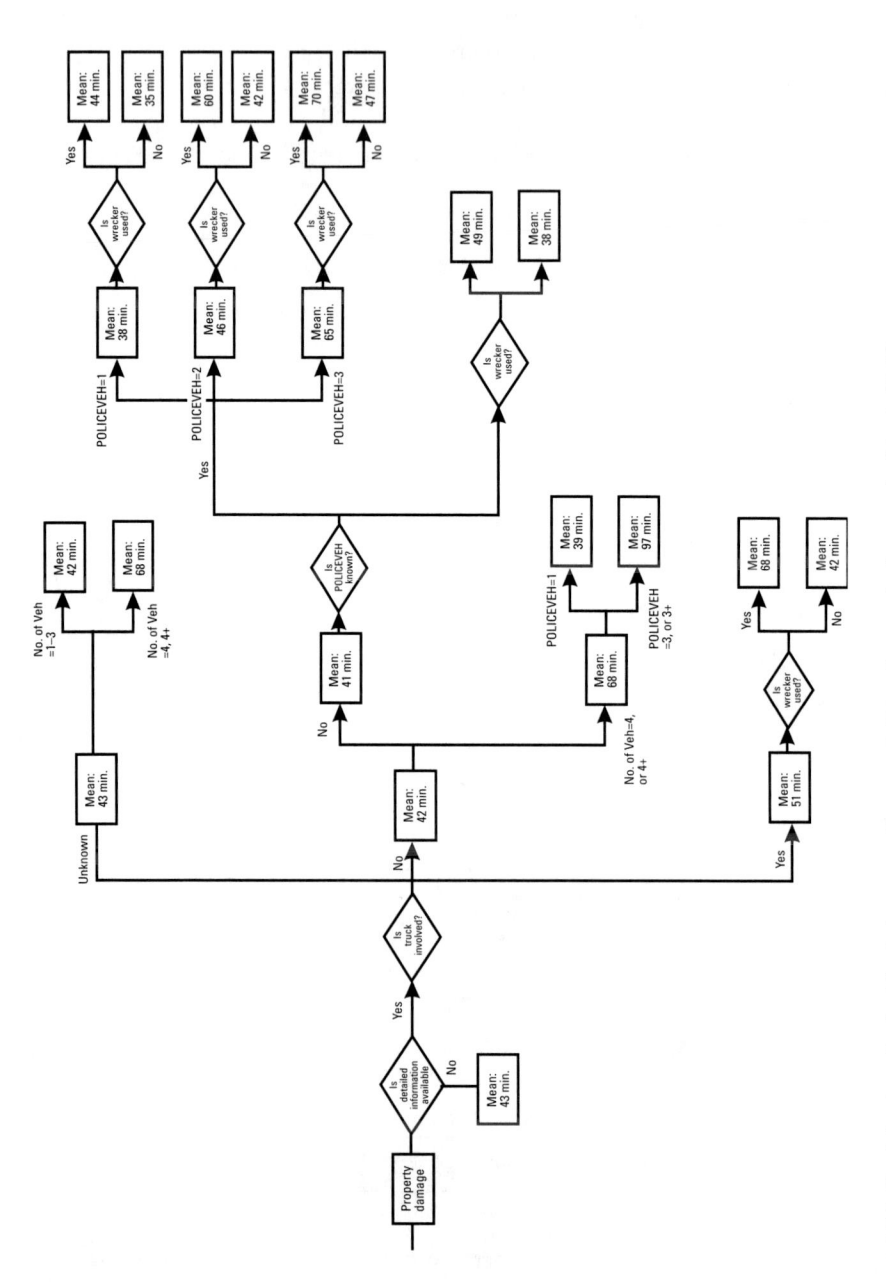

Figure 5.14 Decision tree for incident clearance time prediction of property damage incidents [2, 3].

Incident Duration and Delay Prediction

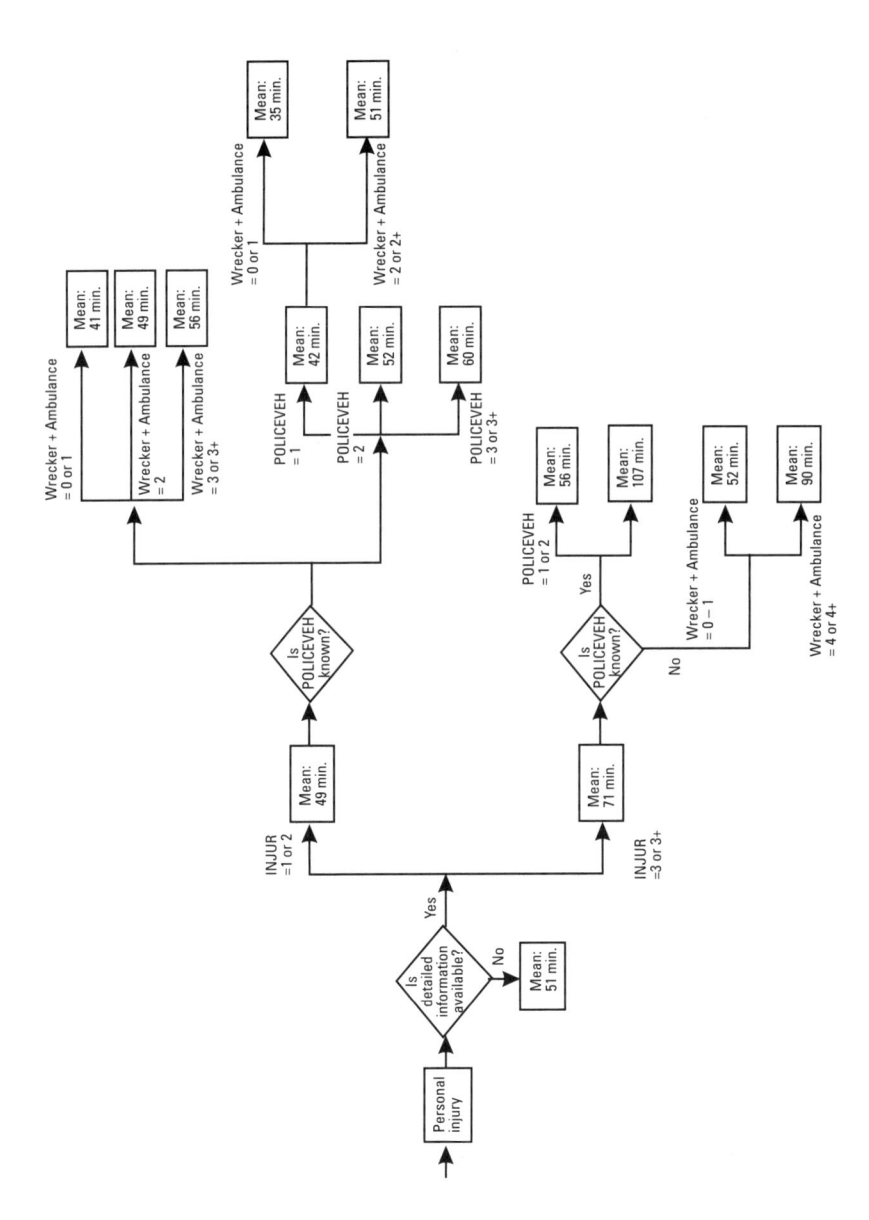

Figure 5.15 Decision tree for incident clearance time prediction of personal injury incidents [2, 3].

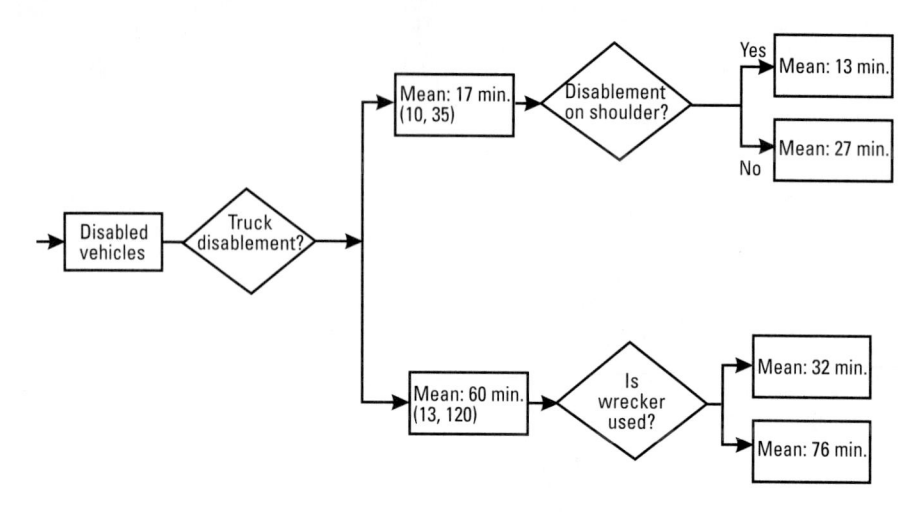

Figure 5.16 Decision tree for incident clearance time prediction of disablement incidents [2, 3].

tunnel. Our data showed that the existence of those factors increased incident clearance times between 12 and 8 minutes. However, sufficient data did not exist to determine quantitative values of those adjustment factors.

It should be pointed out that with more incident data becoming available for analysis and with the help of expert knowledge, decision trees can be refined and further expanded to include more detailed incident types.

5.2.5.1 Disablement Incidents

For disablement incidents, the data came from survey forms of two stages, so both old data and new data were used. Old data were used mainly for disablement on shoulder (no lane closure), while the new data sets are used for more severe disablement cases, namely, disablements in travel lane with lane closure and truck (tractor-trailer) disablements. Figure 5.14 shows the prediction/decision tree developed for disablement incident duration prediction.

It clearly can be seen that truck disablement lasts much longer than normal car disablement, especially if wreckers are needed, in which case the truck disablement can no longer be considered a minor incident. Another interesting point is the significance of lane blockage (closure) on the clearance duration of car disablement incidents. Unlike the cases of property damage, personal injury, and other severe incidents in which lane closure is found to be insignificant with regard to clearance time, lane blockage plays a major role in car disablement. Car disablement occurring on the shoulder deserves little attention and takes little time to clear, while car disablement in the travel lane with lane blockage takes a considerably longer time to clear.

5.2.6 Validation of Prediction/Decision Trees

Data from the new survey forms are used to validate the decision trees. To obtain more convincing results, only the data not used for developing the decision trees are used for validation purposes. Tables 5.6 and 5.7 and Figure 5.17 show the accuracy and the validity of the decision trees developed. It is shown that the developed prediction/decision trees have satisfactory precision in prediction duration of most incident cases: 44 incident cases out of 73 have been predicted with less than 10 minutes of prediction error. For the incident management process, that is an acceptable result, because those predictions are going to be used only as advisories and yardsticks for making diversion decisions. However, some outliers with a large difference between recorded and predicted incident duration exist. The problem is largely due to the individual differences of incident management teams in clearing similar incidents.

Table 5.6
Summary of Validation Results of Prediction/Decision Trees [2, 3]

Differences	Number of Test cases
Predicted CLT-Recorded CLT<=10 min	44 cases
15 min. < Predicted CLT-Recorded CLT<=20 min.	6 cases
10 min. < Predicted CLT-Recorded CLT<=1 min.	14 cases
Predicted CLT-Recorded CLT>30 min.	2 cases
20 < Predicted CLT-Recorded CLT<=10 min.	7 cases
Total No. of Cases	73 cases

Table 5.7
Validation Results [2, 3]

Incident	Predicted	Recorded	Difference
1	35	60	−25
2	44	70	−26
3	35	37	−2
4	35	45	−10

Table 5.7 (continued)

Incident	Predicted	Recorded	Difference
5	38	40	−2
6	46	60	−14
7	52	60	−8
8	52	60	−8
9	35	15	20
10	35	10	25
11	107	110	−3
12	56	40	16
13	35	30	5
14	35	15	20
15	35	20	15
16	51	60	−9
17	35	30	5
18	43	45	−2
19	44	90	−46
20	56	70	−14
21	52	60	−8
22	35	30	5
23	27	10	17
24	35	30	5
25	51	45	6
26	44	30	14
27	51	60	−9
28	35	20	15
29	34	45	−11
30	35	35	0
31	35	35	0
32	35	30	5
33	51	45	6
34	60	60	0
35	44	45	−1
36	52	50	2
37	60	60	0

Table 5.7 (continued)

Incident	Predicted	Recorded	Difference
38	71	60	11
39	44	45	−1
40	44	50	−6
41	44	45	−1
42	35	30	5
43	44	30	14
44	35	45	−10
45	46	45	1
46	35	30	5
47	44	30	14
48	35	30	5
49	44	60	−16
50	44	30	14
51	44	30	14
52	35	40	−5
53	44	30	14
54	60	90	−30
55	35	30	5
56	35	45	−10
57	44	34	10
58	51	48	3
59	107	83	24
60	60	50	10
61	51	70	−19
62	60	59	1
63	35	10	25
64	43	35	8
65	43	30	13
66	76	160	−84
67	76	63	13
68	27	32	−5
69	32	34	−2
70	27	20	7

Table 5.7 (continued)

Incident	Predicted	Recorded	Difference
71	60	35	25
72	27	35	−8

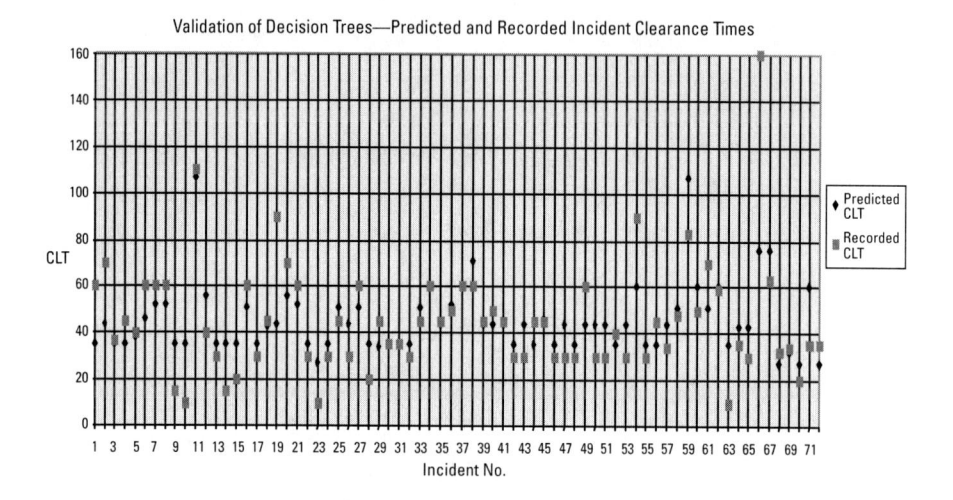

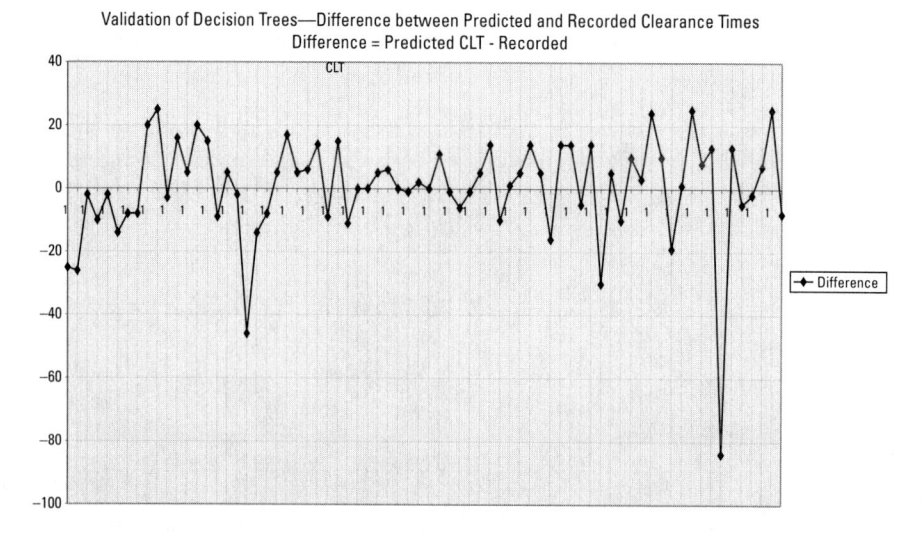

Figure 5.17 Recorded incident clearance times versus those predicted by prediction/decision trees [2, 3].

5.2.7 Distribution Properties of Incident Duration Data Collected for Case Study

The study of incident duration distribution properties provides meaningful assistance in decision making during the incident management process. With the validated incident duration distribution, we can calculate the probability of an incident lasting more than some time period. That information then can be used by the incident management teams for both online and offline studies. One way of using the probability that an incident will last more than some pre-determined value is to decide whether a diversion is needed.

Different researchers studied the distribution properties of incident durations for different types of incidents [4, 5, 14]. To better understand the incident clearance characteristics specific to the northern Virginia area, incident duration distribution properties are also studied in our project. Incident data from a joint survey conducted by Virginia Tech, VDOT, the Virginia State Police, the Fairfax Fire and Rescue Department, and the Fairfax County Police Department in the northern Virginia area are used for this study.

It is found that for a homogeneous subset with enough data samples, the incident clearance time generally conforms to normal distribution. Normal or approximate normal distribution assumptions have been validated by chi-square tests for subcategories of property damage incidents with similar major attribute values (e.g., number of vehicles involved, number of police vehicles on the scene) and other subsets of major incident types.

For the whole data set or data sets classified according to major incident types, the incident clearance time distribution curves shift to the left because the majority of the incidents are less severe ones with short duration. The curves show some trend of log-normal distribution similar to the previous studies by other researchers, but the log-normal distribution assumptions are rejected by statistical tests. However, when the data sets are further divided into smaller subsets based on values in such a way that each subset contains the incident data of the same nature and similar severity, a normal distribution trend becomes apparent and the normal distribution assumption is also confirmed by statistical tests.

It should be pointed out that chi-square tests rather than other higher power tests are used to conduct normality tests in our study. Normal distribution is by nature a continuous distribution, and statistical tests such as the Anderson-Darling generally have high power for testing the normality. However, our duration data are obtained from survey forms and are not recorded very accurately. In most cases, "nice" numbers like 30 minutes are used, even though the actual duration times could be 27, 32, or 33.5 minutes. Therefore, it is appropriate to group the incident duration data into some interval of 15 minutes and to use a chi-square test, which becomes a ready-to-use approach to conduct normality tests. Table 5.8 summarizes the results of normality tests for different groups of incidents.

Table 5.8
Summary of Statistical Test Results for Incident Duration Distribution Properties [2, 3]

Incident Class	Normal Distribution	Log-Normal Distribution	Little Data to Perform Tests
The whole data set	NO	NO	
Road Hazard			✓
Vehicle Fire			✓
HAZMAT			✓
Weather-Related			✓
Cargo Spill			✓
Truck Disablement			✓
Disabled Car in the Travel Lane	YES [$N(27,19^2)$]	NO	
Property Damage, the Whole Data Set	NO	NO	
Property Damage, Trucks Involved	NO	Approximate	
Property Damage, Trucks Involved, Wreckers not Used	YES [$N(42,22^2)$]	NO	
Property Damage, Trucks Involved, Wreckers Used			
Property Damage, No Truck Involved	NO	NO	
Property Damage, No Truck Involved, 1–3 Cars Involved	NO	NO	
Property Damage, No Truck Involved, 1–3 Cars Involved, 1 Police Car on Scene	YES [$N(38,22^2)$]	NO	
Property Damage, No Truck Involved, 1–3 Cars Involved, 2 Police Cars on Scene	YES [$N(46,19^2)$]	NO	
Property Damage, No Truck Involved, 1–3 Cars Involved, 3 or More Police Car on Scene	YES [$N(65,21^2)$]	NO	
Property Damage, No Truck Involved, 4 or More Cars Involved			✓
Personal Injury, the Whole Data Set	Approximate	NO	
Personal Injury, With 1 or 2 Injuries	YES [$N(49,27^2)$]	NO	

Table 5.8 (continued)

Incident Class	Normal Distribution	Log-Normal Distribution	Little Data to Perform Tests
Personal Injury, With 1 or 2 Injuries, 1 Police Car on Scene	YES [N(42,23^2)]	NO	
Personal Injury, With 1 or 2 Injuries, 2 Police Cars on Scene	YES [N(52,28^2)]	NO	
Personal Injury, With 1 or 2 Injuries, 3 or More Police Cars on Scene			✓
Personal Injury, With 3 or More Injuries			✓

5.2.8 Comparison of Our Results With Previous Work

Results of this case study are comparable in many aspects with previous work conducted by [4, 5, 14] and the Northwest ADVANCE Project [8].

5.2.8.1 Incident Characteristics

Both this case study and previous studies have similar conclusions in terms of the percentages of incidents of different severity. The majority of incidents comprise minor incidents, typically disablement incidents. Severe incidents with property damage and personal injuries constitute only a small percentage of the overall data set.

5.2.8.2 Significant Factors Affecting Incident Duration

A Northern Virginia case study results show that truck involvement is a major factor that affects incident duration. Generally, vehicle type is more important than the number of vehicles involved; that result conforms to the results of other studies. Among other attributes, the number of vehicles involved, mainly for nontruck accidents, the number of police vehicles on the scene, emergency equipment on the scene (ambulances, fire engines, wreckers, etc.), and HAZMAT involvement are found to be the significant factors that affect incident duration, which match the results of previous studies. Similar to the conclusions drawn by other studies, inclement weather, rush hour traffic, land use, and roadway type are also found to have some impact on incident duration.

5.2.8.3 Incident Duration

The incident duration is determined mainly by incident type, severity, and clearance characteristics. The different traffic patterns, demand levels, patrol frequency, and response and clearance procedures vary from city to city to make the incident duration somewhat site specific. Nevertheless, the incident durations for accidents of different types in our study are comparable to the results of previous work conducted in other parts of the country (Table 5.9).

5.2.8.4 Variations

Large variations in incident data, making the analysis more difficult and the results less consistent, are observed in this study. Previous studies had similar problems with large variations in data.

The following characteristics distinguish this case study from previous research:

- The main focus of the case study was severe incidents, mainly accidents, represented by property damage and personal injury, while some previous work dealt primarily with minor incidents like ticketing and moving violations. The case study is important for incident management because it focuses on incidents that have a major impact on traffic.

- In previous studies, lane closure was found to be a major significant factor on incident duration and was often used as the main characteristic for incident classification. However, the results of the case study show that it is not a major significant factor for severe incidents.

Table 5.9
Comparison of Incident Duration Obtained in Similar Studies [2, 3]

	Case Study	Giuliano (1989)	Northwest ADVANCE Project(1994)	Goolby & Smith (1971)
Property Damage	43	44	42	45
Personal Injury	51	56	51	*
Disablement	18	*	*	18

* Not Available

In our study, lane closure plays a significant role only in disablement incidents.

- The incident duration distribution conforms to a normal distribution if the data set is homogeneous enough. Log-normal assumptions supported in some studies are rejected in the case study.

- Accident frequency studies conducted in other work are not within the scope of this research project, because the major task of this project is the management processes that occur after the incidents occur.

5.3 Incident Delay Prediction

The second submodule discussed in this chapter is the delay prediction submodule. This submodule adopts the deterministic queuing approach for delay prediction. Delay prediction using the deterministic queuing diagram shown in Figure 5.18 was first proposed by [15]. Different versions of this queuing diagram are proposed by several researchers [7, 16]. This section uses both the basic and the extended versions of this queuing diagram that incorporate the change in traffic demand and freeway capacity to predict the delays. The results of the extended version are also compared with the results of the basic version. An important addition to the delay prediction process is the consideration of the timing of the lane openings. The extended version of the deterministic delay prediction approach incorporates the concept of sequential lane opening as the change in supply or in freeway capacity.

5.3.1 Deterministic Queuing Diagram

This simple model based on deterministic queuing is an analytical procedure that, from the graphs, calculates the cumulative vehicle hours of delay. This method makes a number of assumptions. The first of these is that the demand and capacity are assumed to be constant. The second assumption is that the demand is initially less than capacity. However, this is not always the case, especially in urban areas where many freeways face recurring congestion problem on an everyday basis.

Two of the outputs from the delay estimation procedure are the total time until normal flow (TNF) is resumed, and the total delay in veh-mins, which is the area between the two curves. For the basic version, TNF depends upon several factors.

In [15], Morales gives the term for TNF as follows:

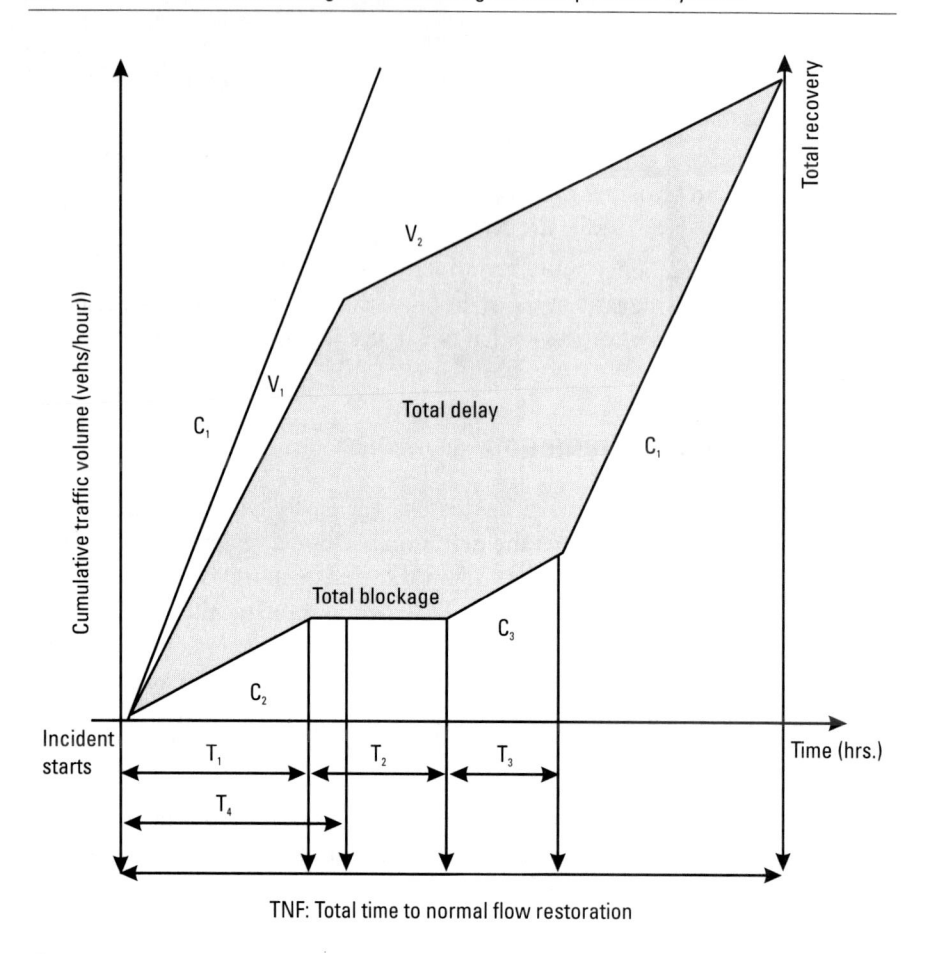

Figure 5.18 Deterministic queuing diagram of vehicle arrivals and departures to freeway section to estimate vehicle delays due to an incident [Based on 7, 15, 16, 17, 18].

$$TNF = \frac{\left[T_1(C_1 - C_2) + T_2 C_1 + T_3(C_1 - C_3) + T_4(V_1 - V_2)\right]}{(C_1 - V_2)}$$

where,
C_1 = Capacity of the freeway (vehs/hr)
C_2 = Reduced capacity due to incident (veh/hr)
C_3 = Adjusted capacity of the freeway (veh/hr)
V_1 = Initial demand on the freeway (veh/hr)
V_2 = Adjusted demand due to diversion (veh/hr)
T_1 = Incident duration until the complete closure (mins)

T_2 = Duration of complete closure (mins)
T_3 = Duration of partial closure (mins)
T_4 = Time until the first change in demand (mins)
TNF = Time to normal flow restoration (mins)

Figure 5.18 employs deterministic queueing concept to illustrate the general relationship between demand (arrivals) and supply (departures) on a freeway section due to an incident. From Figure 5.18, it is clear that the area between the arrival and departure curves is the total incident delay. Morales also gives the following general expression to calculate total delay due to an incident on a freeway [15]:

$$Total\_Delay = \frac{\begin{bmatrix} T_1^2(C_1-C_2)(V_2-C_2) + T_2^2(C_1V_2) + T_3^2(C_1-C_3)(V_2-C_3) \\ +T_4^2(C_1-V_1)(V_1-V_2) + 2T_1T_2C_1(V_2-C_2) + 2T_1T_3(C_1-C_3)(V_2-C_2) \\ +2T_1T_4(C_1-C_2)(V_1-V_2) + 2T_2V_2(C_1-C_3) + 2T_2T_4C_1(V_1-V_2) \\ +2T_3T_4(C_1-C_3)(V_1-V_2) \end{bmatrix}}{2(C_1-V_2)}$$

All of these parameters described above are unknown both at the start and during the progress of an incident. However, they can be estimated by using historical and current traffic data. The variables C_2, C_3, T_2 and T_3 are related to changes in the lane blockage situation, and would have to be adjusted by the system when that information about an incident becomes available. The other three variables, V_1, V_2, and T_4, are related to the change in demand due either to imposed or natural traffic diversions. During an incident with large delays, some of the traffic may avoid the congestion by taking other routes, such as parallel arterials. This decrease in demand is difficult to measure, but should be added as a calibration factor based on historical and existing demands. Morales in [15] also discusses different situations where some occurrences such as demand change due to diversion or complete lane closure, depicted in the general scenario do not happen. It is possible to generate these different scenarios by simple substitution of appropriate values in the TNF and total delay equations shown in this section.

5.3.2 Other Methods to Determine Incident Delays

Although the deterministic queuing method described here is the most widely accepted technique for determining incident delays, it is not as accurate as a simulation-based method. To remedy that problem, some researchers have attempted to develop more accurate incident delay prediction models. The most recent model has been proposed by [16]. They used the I-880 data set to develop regression models to predict the incident delays and have proposed two separate models shown next [16].

Model 1: $Delay = -4.26 + 9.71X_1X_2 + 0.5X_1X_3 + 0.003X_2X_4 + 0.0006X_3$ (5.4)
Model 2: $Delay = -0.288 + 3.8X_1X_2 + 0.51X_1X_3 + 0.06X_3 + 0.356X_2^3$

where

X_1 = number of lanes affected;

X_2 = number of vehicles involved in the incident;

X_3 = incident duration, in minutes (difference between time of incident detection and time of incident clearance);

X_4 = traffic demand upstream of incident 15 minutes before occurrence of incident.

The first model estimates the incident delay as a function of incident duration, prevailing traffic demand, capacity reduction in terms of closed and affected lanes and number of vehicles involved. In fact, this regression model is similar to the mathematical expression of the area between the arrival and departure curves depicted in Figure 5.18. As a matter of fact, it has the same variables of the deterministic queuing model discussed in Section 5.3.1.

The second model developed by [16], which predicts the cumulative incident delay as a function of incident duration, the number of lanes closed or affected, and the number of vehicles involved appears to be a more appropriate model for real-time operations since it does not require a hard-to-get variable, namely, upstream traffic demand. However, the second model incorporates the effect of incident duration on incident delay as a linear one. That, of course, can be considered an unrealistic assumption because it is now widely observed that incident duration is not linearly correlated to incident delay. Another problem with the second model appears to be the incorporation of the cubic number of vehicles involved. In spite of those problems, the first model especially is a practical way of estimating delays due to incidents, particularly for online applications.

5.4 Summary

This chapter discussed an important aspect of incident management process, namely, incident duration/delay estimation. Effective incident management strategies can be developed in real time if and only if the duration of an incident and delay due to the incident can be estimated accurately. Incident management strategies such as establishing alternative routes for traffic diversion around the incident and modifying signal timing plans on neighboring arterials can be developed only if the duration and the delay due to an incident are known with some level of confidence. Such delay and duration information is especially crucial for providing motorists with reliable traveler information. Today, most of the VMSs display information such as "Congestion ahead." That type of information is not very useful to travelers in terms of assessing their decisions for diverting to an alternative route or delaying their trip for some time. To make such decisions, motorists need more detailed information such as expected duration of the incident and expected delay. The duration estimation and delay prediction models presented in this chapter can be used to provide that kind of information to the motorists through different means of information dissemination technologies, including VMS, radio, the Internet, and TV.

It is clear that the decision trees developed for incident duration estimation can be improved and enhanced by continuously calibrating them with new incident data. However, a new study that used an incident data set obtained from Northern Virginia Traffic Control Center strongly supported WAIMSS estimations of incident duration distribution, mean, and variance [16]. The error analysis also provided encouraging results based on the distribution of estimation errors and estimation error percentages. Thus, we can conclude that these WAIMSS duration estimation models perform quite well for the specific area for which they were developed. It is important, however, to emphasize the fact that the average values of incidents have to be calibrated if the same model will be used in other areas of the country.

Review Questions

1. Discuss the variables used to develop incident duration estimation models of the Northern Virginia study. How easy would it be to get this kind of data in your area?

2. Discuss different ways to estimate incident delay. Write a simple computer program to estimate incident durations using the deterministic queuing approach.

3. Develop an incident scenario. Estimate incident delay using the deterministic queuing approach and models developed by [16]. Compare your results. Comment.

4. What kind of traveler information would you give motorists during an incident? Discuss the pros and cons of using duration and delay estimation models to achieve your objective.

References

[1] Subramaniam, S., et al., *Progress Report of Wide-Area Incident Management Project*, Virginia Polytechnic Institute and State University Center for Transportation Research, 1994.

[2] Ozbay, K., A. G. Hobeika, and Y. Zhang, "Estimation of Duration of Incidents in Northern Virginia," presented at the 1997 TRB Annual Conf., Washington, D.C., Preprint #971293, 1997.

[3] Kachroo, P., K. Ozbay, Y. Zhang and W., Wei, "Development of a Wide Area Incident Management Expert System (WAIMSS) Software," (work order #DTFH71-DP86-VA-20), FHWA Final Report, 1997.

[4] Jones, B., L. Janssen, and F. Mannering, "Analysis of the Frequency and Duration of Freeway Accidents in Seattle," *Accident Analysis and Prevention*, Vol. 23, No. 4, 1991, pp. 239–255.

[5] Golob, T. F., W. W. Recker, and J. D. Leonard, "An Analysis of the Severity and Incident Duration of Truck Involved Freeway Accident," *Accident Analysis and Prevention*, 19, 1987, pp. 375–395.

[6] Wang, M. "Modeling Freeway Incident Clearence Times," Unpublished MS thesis, Civil Engineering Dept., Northwestern University, Evanston, IL, 1991.

[7] Khattak, A., J. Shofer, and M. Wang, "A Simple Procedure for Predicting Freeway Incident Duration," *73rd Ann. Meeting Transportation Research Board*, January 1994, Washington, D.C.

[8] Sethi, V., F.S. Koppelman, C.P. Flannery, N. Bhomderi, and J.Schofer, *Duration and Travel Time Impacts of Incidents—ADVANCE Project Technical Report*, Northwestern University, Evanston, IL, 1994.

[9] Sullivan, E.C., "New Model for Predicting Incidents and Incident Delay," ASCE Journal of Transportation Engineering, July-August, 1997, pp. 267-275.

[10] Skabardonis, A., et al., *Freeway Service Patrols Evaluation*, California Path Research Report, UCB-ITS-PRR-95-5, University of California Berkeley, Institute of Transportation Studies, Berkeley, CA, 1995.

[11] Garib, A., A. E. Radwan, and H. Al-Deek, "Estimating Magnitude and Duration of Incident Delays," *ASCE J. Transportation Engineering*, Nov.-Dec. 1997, pp. 459–466.

[12] Petty, K., *FSP 1.1: The Analysis Software for the FSP Project*, University of California Berkeley, Institute of Transportation Studies, Berkeley, CA, 1995.

[13] Breiman, L., J. Friedman, and C. Stone, *Classification and Regression Trees*, Wadsworth & Brooks/Cole Advanced Books & Software, 1984.

[14] Giuliano, G., "Incident Characteristics, Frequency, and Duration on a High Volume Urban Freeway," *Transportation Research—A*, Vol. 23A, No. 5, 1989, pp. 187–396.

[15] Morales, J. M., "Analytical Procedures for Estimating Freeway Traffic Congestion," *Public Roads*, Vol. 50, number 2, September 1986, pp. 55–61.

[16] Al-Deek, H. M., A. M. Garib, and A. E. Radwan, "Methods for Estimating Freeway Incident Congestion, Parts 1 and 2," *74th Ann. Meeting Transportation Research Board*, Preprint #950705, 1995, Washington, D.C.

[17] Freeway Incident Management, NCHRP Synthesis of Highway Practuce, Number 156, 1990.

[18] Morales, J.M., "Analytical Procedures for Estimating Freeway Traffic Congestion," in Traffic Management for Freeway Emergencies and Special Events, Transportation Research Circular, January, 1989, pp. 38-46.

Selected Bibliography

Wei, W., P. Kachroo, and K. Ozbay, "Validation of WAIMSS Incident Duration Estimation Model," to be presented at the 1998 SMC Conf. and published in *Proc. 1998 SMC Conf.*, San Diego, CA, 1998.

6

Incident Response

6.1 The Incident Response Problem

Traffic incident response can be categorized as a subset of emergency response operations that take place in every state. Although emergency response operations can cover events other than traffic accidents and incidents (e.g., natural disasters), this chapter examines only freeway incident response operations, since traffic incidents are the major focus area of this book.

Incident response measures are developed to deal with each incident in an effective and timely manner by deploying appropriate response units to the incident scene and clearing the incident as quickly as possible. It is clear that individual incidents require different types and numbers of response units. The severity and impact of an incident are also important considerations in developing response plans. In general, the expected duration of and expected delay caused by an incident are good measures of the severity of an incident. Thus, the duration and delay estimate models developed in Chapter 5 provide the basis for the development of incident response strategies. In addition to optimally allocating the required resources to clear an incident, information regarding the effect of an incident on traffic is also needed to inform drivers and develop appropriate traffic control strategies, such as traffic diversion.

Traffic incident response can be divided into two major stages. Stage 1 corresponds to all the actions from the verification of the incident to just before the actual beginning of clearance operations. The first stage involves the determination of the incident response agencies required to clear the specific incident, locating the closest incident response agencies, notifying those agencies, coordinating their activities, and possibly suggesting the resources (such as the number of ambulances) required. Stage 2 is basically concerned with various traffic management activities, including information dissemination, ramp metering, posting messages on VMSs, diversion of traffic around the incident and signal retiming. Currently, traffic information, traffic-responsive ramp metering and diversion are the most frequently used traffic management strategies. However, coordinated corridor-wide traffic control strategies that involve, for example, signal retiming on arterial streets, are becoming more common. Some important aspects of traffic management and control activities, such as traffic diversion, ramp metering, and signal timing, are discussed in Chapters 7 and 8.

The following questions need to be answered to understand the complexity of the incident response problem in the context of stage 1:

- How can we determine the optimal number and type of response units such as police cars, fire engines, and wreckers in real time?
- What are the optimal system configurations for locating various incident management agencies and personnel on the network?
- How can the response time be minimized?
- How can we optimally deal with multiple incidents?
- How can incident management agencies and others establish priorities, develop plans, and coordinate various incident management activities in a real-time situation?
- What is the tradeoff between the response time and the monetary cost of minimizing the response time?
- What are the real and perceived benefits and costs of incident management plans that are operational in numerous cities of all sizes in the United States?

This chapter simplifies the problem of incident response to minimizing response time and optimizing resource allocation. The first priority of any incident management program is to clear traffic incidents as quickly as possible to minimize their adverse impacts on traffic flow. According to that approach, which has been proposed in [1], incident response comprises the following four components:

- The incident detection and verification time (T_1) is the time between the actual occurrence of an incident and its detection and verification by the TOC.

- The dispatch time of incident response team(s) (T_2) is the time between the identification of the incident location, type, and severity and the dispatch of the first available incident response team(s).
- The travel time of the incident response team to the incident scene (T_3) is the time required for the incident response team to arrive at the incident site.
- The incident clearance time (T_4) is the time required to fully clear the incident and remove all the involved vehicles and people from the incident site.

Reducing any of those four components clearly reduces the overall response time. Incident management teams throughout the country have long recognized the importance of reducing the response time, because reduced response time means reduced delays. There is also some degree of relationship between the length of time each component takes. For example, incident team travel time to the scene might be longer if the dispatch time is longer, because traffic congestion due to the incident slows down the upstream traffic as a function of time. On the other hand, the efficiency of the incident response primarily depends on the availability of resources needed to clear the incidents. That becomes especially important if more than one incident occurs simultaneously. After an incident has been detected and verified, the scene manager, who in most cases is a police officer, either decides and conveys the requirement of the resources to the TOC or informs the center about the type and severity of the incident so the controller can make a decision regarding allocation of resources.

Incident detection and verification time (T_1) and clearance time (T_4) were discussed in Chapters 4 and 5, respectively. This chapter deals exclusively with the dispatch time (T_2) and the travel time (T_3) of the incident response team. The heuristic rules that will be discussed in this chapter can be used for optimal resource allocation in reducing dispatch delay (T_2) that is due mainly to dispatch of inappropriate equipment.

6.1.1 Tools

According to the National Incident Management Coalition, several tools are available to deal effectively with incident response problems. Some of those tools are the effective organization of response teams, improved interagency communication, use of dedicated highway service patrols, and issuing of tow truck contracts to private operators.

In general, the goal is the optimization of resource requirements by enabling the decision maker to accurately estimate the optimal resource requirements for different types of incidents. That important goal, which is also called optimal resource allocation, can be achieved by providing the decision maker with a set of resource requirements for each incident type (with different attributes) in a manual form or as a computer-based decision support system. In other words, a formal response plan that can deal with different kinds of incident situations should be formulated for each incident category.

The rules of a response plan can be developed with the help of expert knowledge in the form of a rule base that contains the optimum resource requirements for various kinds of incidents.

The expert knowledge used to develop the rule base can be acquired from incident management-related personnel such as highway patrol officers, scene managers, and fire and rescue dispatchers. In the WAIMSS project, a historical incident database is used mainly to determine the resources needed for each incident category. The case study presented in this chapter, which is discussed in detail in [2], summarizes the efforts for developing an expert knowledge base for the efficient identification of resources during real-time incident management process.

6.1.2 Research Needs for the Development of Incident Response Support Tools

Effective and timely incident response remains one of the most important parts of the incident management problem, basically because of its complex nature. Incident response is one of the areas of incident management in which there still is room to make major improvements by providing incident management teams with the necessary tools. Incident response basically is a command-control problem that involves many players. Support tools coupled with advanced communication capabilities can undoubtedly improve overall response time to levels much lower than today's response times. Another important area in which major improvements can be achieved through research is the optimal use of existing resources to handle more incidents. Unlike other aspects of incident response such as dynamic alternative routing, ramp metering, signal retiming, and traveler information dissemination, which received extensive attention from ITS researchers, optimal location of incident response agencies and optimal resource allocation did not receive much attention. However, one of the most efficient ways of reducing adverse impacts of an incident after its detection and verification is through a series of quick and effective response actions.

Because incidents vary in type, severity, location, and time, a host of organizations, each with different responsibilities and capabilities, may be involved in the incident response process. Thus, efforts for coordinating the activities of various agencies after an incident has occurred can be unrealistic and unproductive. It is necessary to develop a set of systematic procedures beforehand. Those procedures then can be integrated in a computer-based decision support tool that will assist with incident management operations. The WAIMSS incident response module consists of simple rules that capture systematic response procedures as part of the overall research efforts at the Virginia Tech Center for Transportation Research [2].

The presentation of those research efforts constitutes the rest of this chapter. First, the existing incident response plan development efforts throughout the United States are presented. Then, a comprehensive incident response methodology that can be used to improve existing incident response efforts is proposed. The implementation of that methodology within the framework of WAIMSS software developed at the Virginia Tech Center for Transportation Research (discussed in Chapter 3) is also presented. Like the rest of the WAIMSS, the response rules developed using the database are implemented with the expert system shell Nexpert-Object. The GIS capabilities of WAIMSS have been used not only to display the network but also to analyze the spatial relationships between incidents and response centers. The GIS component of WAIMSS also can be used to minimize the travel time (T_3) by quickly identifying the best response locations that have the adequate equipment for responding to a specific incident and determining the fastest route from the response center to the incident site.

Finally, a case study analyzes the incident response operations in northern Virginia using the incident data collected by the NOVA incident management personnel and researchers at Virginia Tech Center for Transportation Research.

6.2 Existing Incident Response Systems

Several important resources describe actual incident response activities in various states: the operating manual for the Northern Virginia Freeway Management Team [3], the final report for the Caltrans District 12 Real-Time Expert System for Freeway Incident Management [4], and the state-of-the-practice reports for the I-95 Corridor Coalition [5, 6]. Various other papers also have been studied to better understand the existing response programs [1, 7–10].

6.2.1 Orange County, California: Caltrans

The Caltrans-Orange County incident management program is one of the leading incident management programs in the nation. A real-time expert system for efficient response to freeway incidents has been developed by the

University of California, Irvine (UCI) for Caltrans District 12 (Orange County). The overall incident management plan currently being utilized in Caltrans District 12 is shown in Figure 6.1.

After meeting with the experts over a period of time, the researchers at UCI came up with the following incident types [4]: vehicle fire, traffic collision, vehicle fuel spill, bomb threat, crime investigation, planned lane closure, HAZMAT fire, stalled vehicle, gawking, HAZMAT, overturned truck, spilled load, jackknifed truck, suicide threat, special event, emergency lane closure, adjacent fire, load fire, and fog/dust/smoke.

In this study [3], incidents are classified as low, moderate, and high-impact incidents depending on their severity and the time of occurrence. After discussing various parameters such as jurisdictions, responsibilities of agencies, and incident characterization, the UCI researchers formulated a response plan. They presented the response plan implemented using the G2 expert system shell to the operator at the District 12 control center. The developed computer-based tool interacts through the system map; an incident summary, which is completed by the operator on the basis of the information received; an incident checklist, which is built into the system and can be changed over time; and the detailed specific response window, which has the standard responses in the order that the operator needs to address them.

The responses are as follows [4]:

- Issue "Sigalert."
- Issue traffic advisory.
- Advise traffic management team (TMT) team leader.
- Alert Caltrans organization.
- Activate CMSs.
- Advise maintenance.
- Dispatch FSP/Orange Angels.
- Contact TMC supervisor.
- Notify duty officer.
- Put incident on California Highway Information Network (CHIN).
- Advise local agencies.
- Advise the adjacent district.
- Check HAR.
- Investigate using the CHP helicopter.
- Close ramps.

Incident Response

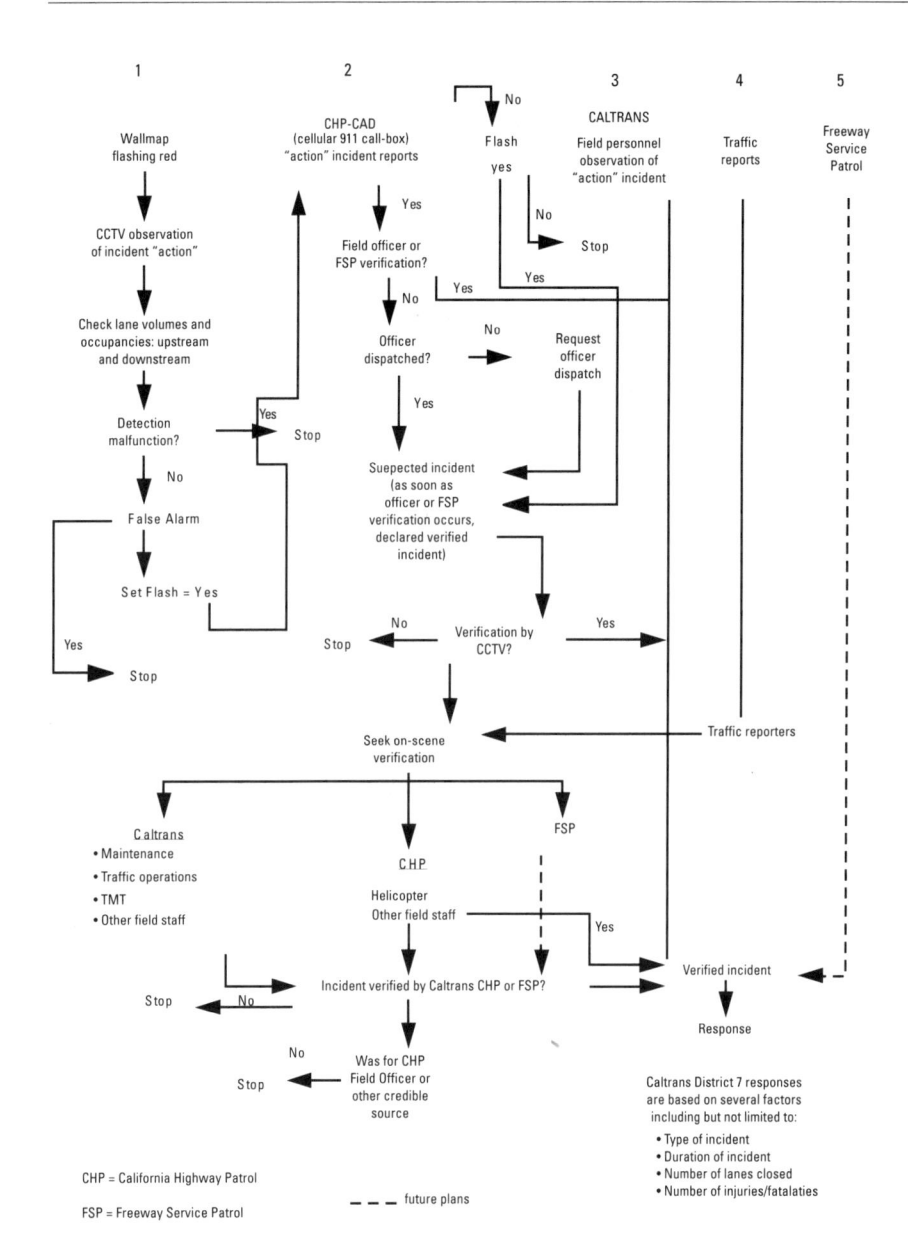

Figure 6.1 Incident management plan being followed at Caltrans District 12 [4]. (*Source*: S.G Stack and G. R. Ritchie, University of California, Irvine, 1995. Reprinted with permission.)

A typical statement in the response plan devised for the expert system is shown in Figure 6.2.

```
         Notify              IF notification of a potential Major Incident
         TOC Supervisor      THEN advise TOC supervisor

         Dispatch TMT        IF confirmed
                             THEN dispatch TMT
         Maintenance         IF confirmed incident
                             THEN notify Maintenance of the Incident
```

Figure 6.2 Response module statements in the expert system developed by Stack et al. [4] (*Source*: S.G. Stack and G. R. Ritchie, University of California, Irvine, 1995. Reprinted with permission.)

The developed system is an easy-to-use and comprehensive tool that can provide the operators with useful information for developing response plans.

6.2.2 I-95 Corridor Coalition

The I-95 Corridor Coalition comprises the northeastern states of Maine, Vermont, New Hampshire, Massachusetts, Connecticut, Rhode Island, Pennsylvania, New Jersey, Maryland, Delaware, and Virginia. The coalition has the goal of improving the organization of the incident management programs in these states and coordinating the tasks of various agencies throughout the corridor to improve the incident management process. The agencies include the DOTs, turnpike authorities (TAs), highway authorities (HAs), departments of public works (DPWs), and other private agencies of those states [5, 6].

Once the incident is verified, the response procedures commence. Table 6.1 identifies some of the responses by coalition and state agencies relative to key incident response issues. Response is defined as "the activation of a preplanned strategy to minimize the adverse impacts of the incident." The effective response requirements are defined as:

- Determining the nature and location;
- Initiating the appropriate plan;
- First response;
- Monitoring, evaluating, and adjusting;
- Training and practice.

Table 6.1
Incident Response Procedures in the I-95 Corridor [5-6]
(With permission from : I-95 Corridor Coalition, 1995.)

Issues	VDOT	PennDOT	NJDOT	California	NYSDOT
Process for Incident Response	Y	N	Y	Y	N
Procedures for Multi-Agency Involvement	Y	N	Y	N	Y
Who defines the Incident Response Levels? (Legend A)	1,3	1	1,2,3	1	1
Who is responsible for (Legend B):					
Fatalities	P	C,P	C,E,P	C	P
Injuries	E,F	E	E,P	E	P
Vehicle Contents	P	P	P,T	P	P
Damaged Vehicles	P	P	P,T	P	P
Leakage	F,H	P	D,H	D	F
Traffic Control	D,P	P	D,P	D,P	P
Traffic Flow	P	P	P	D	P
Who is the responsible person on the scene for: (Legend B)					
non-HAZMAT	P	P	P	P	P
HAZMAT	F	P	H	P	F
Must Coroner respond to the scene in case of fatalities?	N	Y	N	N	Y
Means to obtain equipment to clear incidents? (Legend C)	3	2	2	3	2

LEGEND A
    1—Police officer at scene, 2—Agency Personnel at scene, 3—Fire Personnel at scene
LEGEND B
    C—Coroner/Medical Examiner, D—DOT/Agency, E—Emergency Medical Services, F—Fire Department,
    H—HAZMAT/Department of Environmental Protection(DEP)/Department of Environmental Resources
    (DER)/State Emergency Response, P—Police, T—Tow Service Provider
LEGEND C
    1—Utilization of Agency resources, 2—Utilization of private services, 3—Combination of Agency/
    private services

The I-95 Corridor Coalition identifies incident response as a key component of the incident management process. According to the report by the I-95 Corrdior Coalition [5], where in some cases, the state highway agencies do not have a list of procedures to be followed in response to an incident, the level of response typically is determined by a duty officer, shift supervisor, or dispatcher at the communications center. Some agencies coordinate the dispatch activities by housing the police dispatch in the same building as the highway department.

The resources required for the response usually are identified by a preplanned call list of responders. While most incident management programs on the I-95 Corridor are based on typed lists and procedures, several highway agencies are in the process of developing automated processes to reduce response times.

The coordination among various agencies is identified as a key issue. After the incident occurs, all agencies must work in coordination to complete the selected response procedure according to a plan that clearly identifies the response procedures. To accomplish that task without causing any delays while the individual responsibilities of each agency are being decided, all participating agencies have to be specified in advance.

Once the incident response procedures are in place, management of traffic to move it through the incident site in a safe and efficient manner becomes a key issue. Two of the most important traffic management issues are rapid clearance of the incident and restoration of normal traffic conditions. Certain agencies in the coalition have quick clearance policies that involve clearance of the incident as soon as possible. As part of those quick clearance policies, the incident management agencies, including tow operators, fire departments, police departments, and DOTs are instructed to minimize the blockage of lanes at the incident site. The area is physically screened off by parking vehicles and placing equipment at advantageous locations to reduce the gawking effect [5]. Table 6.2 shows various traffic management procedures. The overall goal is to minimize the disruption to traffic. For example, the VDOT has an abandoned-vehicle policy, which permits abandoned vehicles to be removed from the road after 24 hours. The VDOT does not have a hold-harmless policy, which permits small incidents to be cleared in minimum time by risking small damage at the cost of the owner or the insurance company [2, 5].

Notification of drivers about an incident and traffic diversions is another important task addressed by the I-95 corridor coalition, usually done through radio broadcasts and VMSs. The I-95 corridor has independent traffic reporting agencies such as Metro Traffic and Shadow Traffic. HAR is also a good alternative means of reporting incident-related data.

Table 6.2
Traffic management procedures in the I-95 Corridor [5-6]
(With permision from: I-95 Corridor Coalition, 1995.)

Issues	VDOT	PennDOT	NJDOT	California	NYSDOT
Role of DOT (Legend A)	1	2	2	1	2
Importance of clearing lanes (Legend B)	3	2/3	3	3	2
Methods used to minimize traffic disruptions					
Minimum lane blockage	Yes	Yes	Yes	Yes	Yes
Vehicle screening					
Physical screens					
Motorist Notification by					
Public Radio	Yes	Yes	Yes	Yes	Yes
HAR	Yes		Yes		
Portable VMS	Yes		Yes		
Fixed VMS	Yes	Yes	Yes	Yes	Yes
Traffic Detour		Yes	Yes		
Pre-Trip Notification	Yes	Yes	Yes	Yes	Yes
Diversion Routing (Legend C)	1	1	3	1	2
When is traffic management terminated ? (Legend D)	2	1	1	2	1

LEGEND A
  1—Advice IM Team Responder, 2—Support Role in IM
LEGEND B
  1—Not important, 2—Less important than property damage, 3—Less important than personal injury,
  4—Most important
LEGEND C
  1—Formal plans through detour, 2—Planned detour off system, 3—Plans not documented, 4—No plans
LEGEND D
  1—Upon completion of on-site clean-up activities, 2—Upon complete clearing of any backups

Throughout the corridor, service patrols are used for incident detection and verification. The primary objective of the highway patrols is to minimize the duration of an incident by its fast removal and to reduce the risks to motorists and response personnel. Table 6.3 shows the service patrols and characteristics particular to Virginia.

Table 6.3
Service patrols and characteristics on I-95 in Virginia. [5, 6]
(With permission from: I-95 Corridor Coalition, 1995.)

Operator	Route	From	To	Hours of Operation
VDOT Northern Virginia District	I-95 I-395 I-66 I-495 Dulles Toll Road American Legion Bridge Woodrow Wilson Bridge	I-495 DC Line DC Line Rt 860 Arlington	Exit 148 I-495 Fauquier Co. Line W.Wilson Bridge Dulles Airport	24 hrs/day 365 days/yr
VDOT Fredricksburg District	I-95	Exit 148	Exit 98	Holiday weekends only 5:30 am to 10:30 pm
Virginia State Police, Richmond District	I-95 I-64 I-195	Richmond City Henrico City Richmond City	Henrico County	weekdays 6am-8pm weekends 10 am-8pm holidays 10 am-9pm
VDOT Suffolk District	Route 44 I-264 I-564 I-64	Toll Plaza Merrimac Blvd Terminal Blvd Bay Ave.	I-64 I-64 I-64 Indian River Rd	24 hrs/day 365 days/yr

[1] Vehicle Types: (a) van, (b) modified pickup truck, (c) wrecker, (d) sedan
[2] Functions: (a) rove and detect, (b) on-call stationary, (c) report to other responders, (d) tow, (e) removal from travel lanes, (f) request tow, (g) assist pedestrians, (h) remove debris, (i) extinguish small fires, (j) provide fuel or other fluids, (k) repair/replace flat tires, (l) minor mechanical repairs, (m) push bumper, (n) two-way radio, (o) cell phone

Most agencies have specific operating guidelines for towing and wrecker services, which are a key component of the incident response plan. The guidelines include criteria such as response times, proximity to the area, and communication access.

6.2.3 Northern Virginia

Based on [3], the Northern Virginia District Transportation Operations Center covers VDOT activities in Arlington, Alexandria, Fairfax, Loudoun, and

Prince William counties. Currently, the center supplies incident detection and verification information to the transportation operations center for action. The transportation operations center provides intra-agency and interagency notification and coordination.

In most instances, the police/fire dispatch center is notified of an incident before anyone else. They receive the notice of an incident via cellular or land-based telephone, from police/fire units on patrol, or from call-for-service. VDOT safety service patrollers make initial notifications of incidents they observe, and the VDOT Traffic Management System may observe an incident via their CCTV cameras.

The Virginia State Police (VSP) is responsible for responding to incidents in the northern Virginia area, with a sergeant responsible for each county in the area. An important component of the incident response efforts in northern Virginia is the safety service patrols, which ensure quick response to traffic incidents. In Virginia, there have been extensive efforts to developing statewide criteria for expansion of safety service patrols and developing a classification system for towing and recovery operations. The classification system is designed to ensure the right dispatch of towing equipment to expedite clearance of roadways [11].

6.2.4 Research on Incident Response

In addition to the real-world projects dealing with incident response, Zografos and colleagues have conducted research in emergency response fleet operations for traffic incident response [1]. The main problem they studied is the spatial management of distributed response services (police, fire, medical services) to determine the number of emergency response units (ERUs) that should be made available [1]. The primary MOE is the minimization of the response time for an ERU. According to [1], a crucial parameter involved in the design and evaluation of emergency response operations is the total incident service time (TIST). TIST is equivalent to the incident response time concept. For a complete discussion of that important paper [1], four distinct parts of TIST are defined [1]:

- *Incident detection and identification time* (T_1). Time elapsed between occurrence of the incident and the arrival of a call at the dispatching center regarding the occurrence of the incident. T_1 can be reduced drastically if the smart-car technology (which involves automatic notification of the police using GPS in case of a collision) replaces current methods.

- *Dispatch delay* (T_2). Time elapsed between the detection and identification of the incident and its assignment to the first available ERU.
- *Travel time* (T_3). Time required by the ERU to travel to the incident site from the resource center.
- *Response activity time* (T_4). Time required to provide the required service at the incident site.

Note that those times are almost identical to the ones presented at the beginning of this chapter. A simulation model for evaluating operations of an emergency fleet was also discussed in [1]. Alternative dispatching strategies, such as first come, first served (FCFS) and nearest neighbor (NN), were simulated, and alternative dispatch policies under various conditions of temporal, spatial, and severity distribution of service calls were evaluated.

6.3 Formulation of a Response Plan

The review of incident response literature is a crucial step in understanding the overall response problem. The rest of this chapter presents the formulation of a response plan that can be used within the framework of WAIMSS and that is based on the research findings also presented in [2, 12]. In the process of the formulation of a response plan, all the incident types have first been broadly classified as *vehicle* and *non-vehicle* incidents, each of which has several subclasses. That allows identification of all services required at the incident site based on the incident's characteristics. The services are then related to the agencies. A list of tasks performed by each agency and interagency responsibilities are then identified.

In the formulation of an appropriate incident response plan, it is also important to capture the process of incident response in the TOC. The first step of the process at the TOC is the creation of an incident report as soon as the incident is detected and verified. The report is then updated throughout the response process to reflect any change in resource requirements and related estimated clearance times. Hence, the incident response process as defined in WAIMSS consists of the following steps:

1. Incident characterization;
2. Service identification;
3. Agency notification;
4. Clearance activity.

That process is illustrated in Figure 6.3. The process described in this figure is iterative because during the clearance process, new incident attributes might be identified by the agencies on the scene. For example, a HAZMAT leak that did not exist originally may be observed at some stage of the incident response process. That information is then sent to the TOC, and the original incident report is revised. That restarts the characterization–service identification–agency notification loop immediately.

6.3.1 Incident Characterization

An incident can be defined as an event that causes blockage of traffic lanes or any kind of restriction of the free movement of traffic. Hence, incidents belong to a wide variety of events, ranging from "accidents" (which may involve collision) to weather-related incidents, such as fog or snowfall. For characterizing an incident, all incidents first have to be classified.

After classifying the incident, the incident attributes have to be identified. Incident attribute information includes any information about the incident that would help in the response process. The information can be stored in a database and used in the future for understanding the response needs for various incidents. In this section, the classification of incidents differs from the classification of incidents presented in Chapter 5, which discussed the duration estimation problem. The incident classifications given here are based on the nature of the response problem, because some of the incident classifications

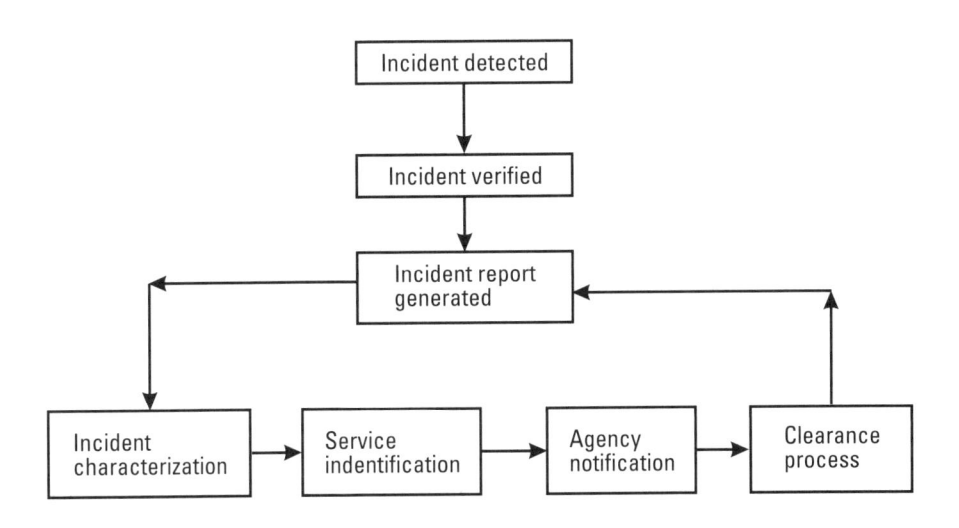

Figure 6.3 Incident response process [2, 12].

have been specifically modified for the response problem. However, WAIMSS has been developed as independent modules, and those kinds of minor differences do not adversely affect the working of the overall system. Various incident attributes defined for the response plan development are as follows:

- *Location.* The location of an incident affects the response process in two ways. The priority of clearing an incident varies with the location of the incident. An incident on a freeway or a bridge, in most cases, would have a higher priority than one on a local route. Second, the location of the resource center from which the resources are to be dispatched depends on the location of the incident.
- *Time of occurrence.* The time of occurrence of an incident affects the response priority significantly. If an incident occurs during or will last into peak hours, a higher priority is given to the incident. The presence of personnel and equipment at the response centers also depends on the time of day and the day of the week.
- *Time of detection.* This information is useful in keeping track of the duration of the incident.
- *Vehicles involved.* The number of cars and/or trucks involved in an incident helps to estimate the resource requirement (discussed in detail in Chapter 5).
- *Injuries.* The number and severity of injuries also reflect the severity and magnitude of an incident and is usually the primary factor in determining the number and type of ambulances that should be dispatched.
- *Fatalities.* The number of fatalities also reflects the severity and magnitude of an incident. In the case of a fatality, the announcement of death by the coroner is required in most states before removal of the body. That, in turn, could adversely affect incident duration.
- *Lanes blocked.* The number of lanes blocked helps to determine the expected traffic delay. Delay is the key in making decisions regarding diversion since one of the most efficient ways of reducing delay is decreasing demand by diverting traffic around the incident.
- *Property damage.* Police investigation is usually required in cases of property damage incidents that involve damage to the vehicle. Incidents with damage to the infrastructure require repair by the DPW.
- *Weather conditions.* Weather conditions affect the efficiency of response, and inclement weather (such as snow) usually imposes special requirements for resources.

- *Cargo spill.* Special equipment is required in incidents involving a cargo spill. The type and amount of equipment dispatched depend on the amount and type of the spill.
- *HAZMAT spill.* Incidents involving HAZMAT spills are responded to by a HAZMAT clearance agency. The whole area might have to be cordoned off if the spill is hazardous to human life.
- *Vehicle breakdown.* A disabled vehicle, in most cases, has to be towed away. That requires wrecker(s) or tow truck(s) at the scene.
- *Vehicle fire.* A vehicle may catch fire in a collision or a noncollision accident. The number and type of vehicles on fire determine the number of fire engines deployed.

6.3.2 Service Identification

An incident is characterized as a specific type so its most significant attributes from the traffic operator's viewpoint can be identified. Characterization of an incident as a specific type helps the operators at the TOC to efficiently collect all the important incident features and enables them to further determine a complete package of services required to alleviate the adverse effects of the incident. Each feature of the incident leads to the requirement of one or many services. The following list is a compilation of possible incident response services performed by various incident management personnel in the United States and identified as a result of a literature survey [3–6]:

- *Incident site management.* Major incidents require the involvement of multiple agencies. In such situations, a site command post has to be established to coordinate activities of different agencies, to manage the incident area, and to collect and disseminate traffic information to travelers. In many areas of the country, the appropriate police department usually undertakes the task of site management.
- *Police investigation.* Most incidents require a police investigation of the incident scene by state or local police or the highway patrol, depending on the location of the incident.
- *Disabled-vehicle removal.* Different types of tow trucks, depending on the type of disabled vehicles, are deployed to remove disabled vehicles from the incident site.
- *Traffic control and management.* Traffic control and management measures include setting up VMSs to provide incident information or to divert traffic, manually directing traffic to alternative diversion

routes, changing ramp metering rates, and re-timing signals on the surface streets.

- *Traffic information dissemination.* The traffic control center informs drivers about traffic conditions and any relevant traffic control measures in place (e.g., alternative routes) and expected delays. *Traffic information dissemination.* The traffic control center informs drivers about traffic conditions and any relevant traffic control measures in place (e.g., alternative routes) and expected delays.
- *Emergency medical services.* Emergency medical services include on-scene treatment and transporting the injured to a medical center using MEDVAC helicopters. In the case of an incident involving fatalities, the coroner must investigate before a victim's body can be removed.
- *Fire extinguishing.* Fire extinguishing is one of the most important operations and is assigned higher priority over most other operations.
- *Spilled material and HAZMAT removal and handling.* Depending on the type of spill, clearance of the roadway is done to ensure safe passage for traffic. Some incidents involve HAZMAT vehicles. In the case of a HAZMAT leak or spill, the problem has to be addressed by personnel with special training and special equipment, usually a professional HAZMAT agency and the relevant environmental protection agency.
- *Road damage repair.* Road damage has to be taken care of immediately by appropriate maintenance and construction crews to avoid a hazard to traffic and a reduction of roadway capacity.

6.3.3 Agency Notification

After the appropriate services have been identified for each incident, the agencies responsible for performing each service must be notified. An effective incident response system must be designed to coordinate service activities of a number of agencies and to minimize the adverse effects of incidents by reducing their response times. Agencies differ in their responsibilities, capabilities, and location, and no single agency can handle all incidents. A major factor in ensuring a smooth and fast incident response operation is the effective coordination of agencies' activities. The following major agencies respond to incidents:

- TMC or traffic control center;
- Local or state police department(s);
- DOT;

Incident Response 151

- Highway patrol;
- Fire department;
- Medical agencies, hospitals, and MEDVAC services (usually helicopter);
- HAZMAT and environmental agencies;
- Towing/wrecking agencies;
- Military;
- Media;
- Department of Public Works (DPW).

Among those agencies, the police department usually is in charge of isolating the incident, controlling people and traffic at the scene, identifying and implementing alternative routing plans, and conducting an accident investigation. If fire is involved, the fire department extinguishes the fire, identifies hazardous materials and any area requiring evacuation, requests clean-up of resources, if necessary, and helps the injured. The DOT provides support material and resources, estimates damage to the infrastructure, and collects data. Recently, the DOT has been playing an integral role in managing traffic in coordination with police, disseminating traffic information using VMSs and HAR, and coordinating multiple-agency activities with the help of a command-level person usually located at the traffic control center to coordinate activities. The DOT also organizes activities to train incident management personnel, to increase preparedness for responding to incidents.

6.3.4 Clearance Process

The clearance process is concerned with any effort at the incident site that helps to clear the roadway for the normal flow of traffic. That task might involve simply moving involved vehicles from the travel lanes or the incident site, getting a trapped passenger out of a car, cleaning a cargo spill, or evacuating injured passengers.

The coordination of agencies during the clearance process is key to the success of the overall operation. The proper coordination of agencies is achieved by the use of predetermined interagency procedures that identify specific tasks for each agency for different incident scenarios. The main goal of interagency coordination is to clear the incident as soon as possible and to avoid duplication or omission of certain clearance activities.

152 Incident Management in Intelligent Transportation Systems

Throughout the process, the TMO plays the role of a central coordinator by organizing the activities of various agencies. It updates the incident report at certain time intervals to reflect additional resources that are needed. Based on the incident response process, an incident response procedure flowchart that describes the overall response procedure was developed and is shown in Figure 6.4. The conceptual procedure chart can be followed after an incident.

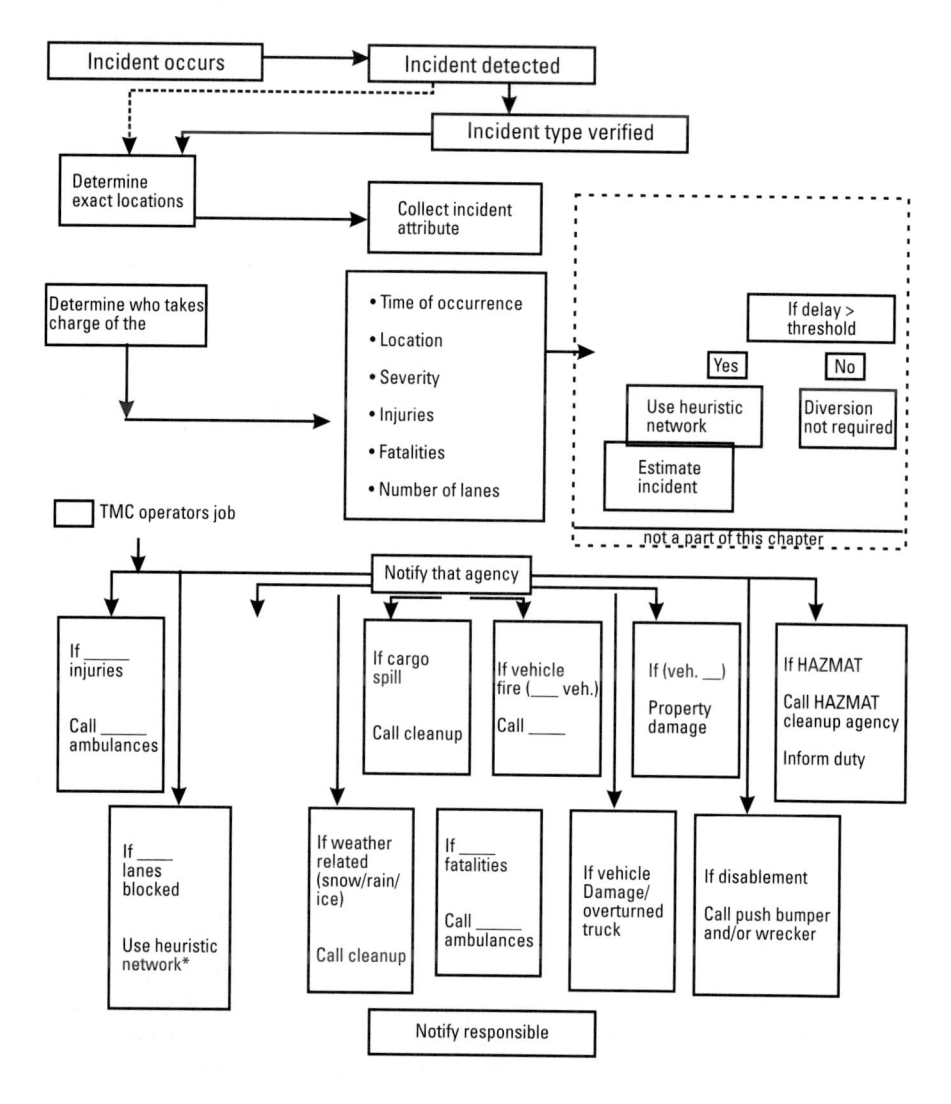

Figure 6.4 Incident response procedure flowchart [2, 12].

6.3.5 Computer Implementation of the Conceptual Computer-Based Response Plan

In Section 6.3.4, a conceptual response plan was formulated. Such a plan can be implemented on a computer for faster decision making. This response plan will be implemented as a GIS (Arc/Info)-based application with minor changes to fit it to the overall framework of WAIMSS. Figure 6.5 shows the implementation of the computer-based response plan, based on the work also presented in [2, 12].

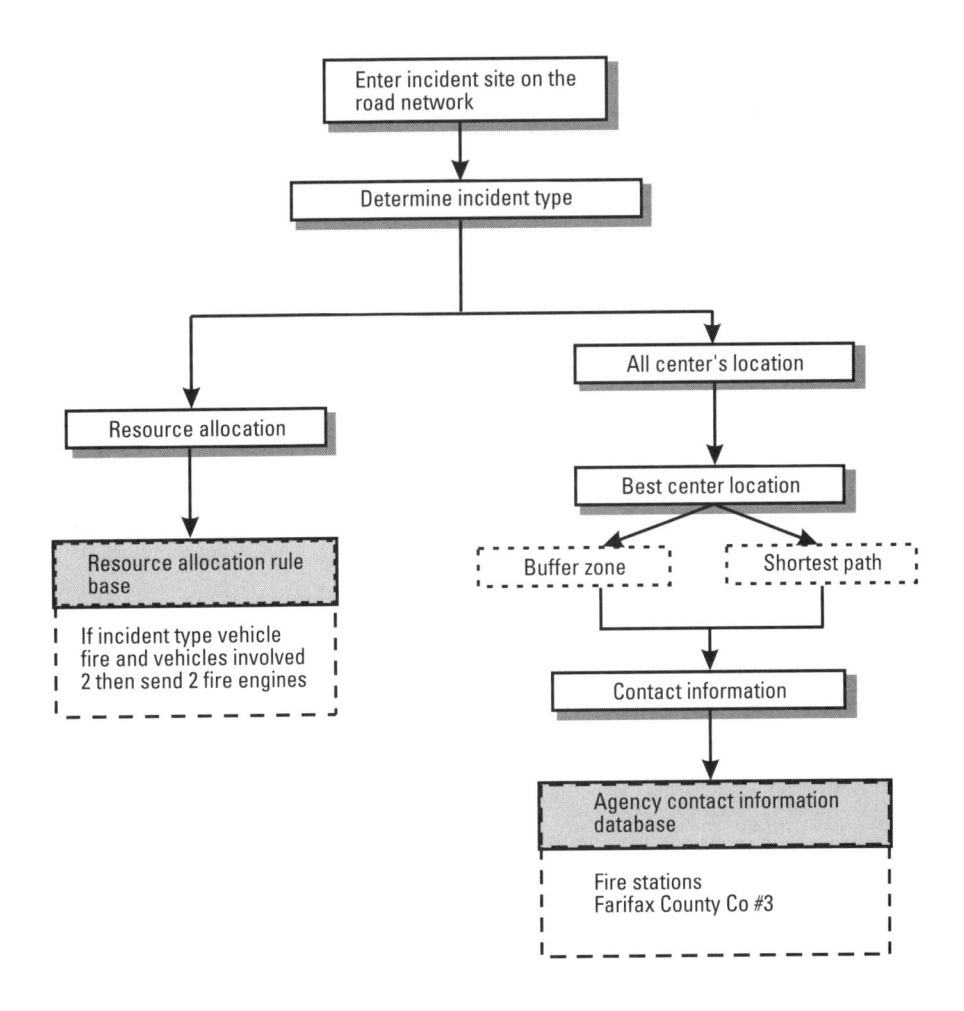

Figure 6.5 Computer implementation of the response plan generation procedure [2, 12].

6.4 Case Study

This section presents the case study conducted at the Virginia Tech Center for Transportation Research using the incident data collected for the WAIMSS project [12]. The goal of this case study is to determine the resources needed to clear incidents in the northern Virginia area. The findings of this case study also are presented in detail in [2].

6.4.1 Study Area and Response Statistics

The study area of the WAIMSS project is Fairfax County, in northern Virginia. The same area is used for conducting studies related to incident response. Due to its proximity to Washington, D.C., the area is highly urbanized and has heavy traffic flow. Recurring as well as nonrecurring congestion are common events and are not limited to the arterial roads. The traffic control center in Fairfax City is responsible for the smooth flow of traffic and dissemination of information in Fairfax County. The police and fire headquarters are located in Fairfax City. There are numerous fire stations, hospitals, and police stations all over the county [12].

As explained in detail in Chapter 5, at the beginning of the WAIMSS project, incident survey forms were sent to police officials at the Fairfax County Police Department (FFCP), the Virginia State Police (VSP), the Highway Patrol, and VDOT. After each incident (except disabled cars on shoulders), the police official at the site filled out the survey form. The forms were mailed to the Virginia Tech Center. All the information has been utilized for various aspects in the project. However, for the incident response portion of the research project, the sections Incident Type, Incident Detail, and Resources Used for Clearance on the survey form were used.

6.4.2 Statistical Analysis of Resources

The aim of the data analysis is to determine the resource requirement for an incident based on the incident type and certain incident attributes which can help develop simple rules for optimal resource allocation in the study area. An important point to emphasize is the meaning of *optimal*. We know that the traffic operators and decision makers at the TMC are experienced professionals who over the years have developed a close-to-optimal decision-making process for resource allocation. Thus, their knowledge for resource allocation can be used as the optimal one. It is clear that different mathematical modeling techniques can also be used to optimize the resource allocation. But due to the specific operational and locational characteristics of this problem, there is no guarantee that new resource allocation guidelines developed using well-known optimization techniques would be the optimal ones. Instead, we have decided to rely solely on expert knowledge and historical data to capture valuable domain knowledge. Simple resource allocation rules based on expert knowledge and supporting historical data that determine the resources required on a day-to-day basis are used to develop the rules. The "optimal resource allocation" rule base of WAIMSS is simply a decision support tool that captures the already existing expertise of incident management personnel in northern Virginia. The approach of using expert knowledge is consistent with the development philosophy of expert systems.

The analysis of the historical incident database was done using Structured Query Language (SQL), available in Fox Pro and Microsoft Excel, to generate the charts presented in this chapter. The analysis based on phase 2 data collection efforts gathered information on major incidents to determine resources used for different types of incidents. The information is then used to develop the resource allocation rule base of WAIMSS. The following subsections present a summary of the analyses also presented in [2, 12].

6.4.2.1 Analysis With Respect to Response Units

First, the minimum number, maximum number, and average number of response units (wreckers, Table 6.4; fire engines, Table 6.5; ambulances, Table 6.6; arrowboards, Table 6.7; and police vehicles, Table 6.8) have been calculated from the database. The tables provide a good idea of the resources required for each incident type, but in some cases the data might not give very good representation of what actually happens in the field due to the small sample size for some incident types or possible errors during data entry.

6.4.2.2 Analysis with Respect to Incident Type

The averages of the response units are calculated for two incident types: personal injury incidents and property damage incidents. That is because of the relatively large sample size for such incidents and because such incidents occur more often than other major incidents.

In Table 6.9 and Figures 6.6, 6.7, and 6.8, a clear trend can be observed. As the average number of injuries in an incident increases, the number of response units also steadily increases. As the number of cars involved increases, the response units required also increases.

6.4.3 Resource Allocation

Optimal resource allocation can avoid waste and duplication of resources. It can also help reduce decision-making time by ensuring quick clearance of

Table 6.4
Number of Wreckers Required for All Incidents [2, 12]

Wreckers

Number	Incident Type	Number of Incidents	Minimum	Maximum	Average
1	Road Hazard	24	0	2	0.21
2	Property Damage	438	0	3	0.37
3	Personal Injury	250	0	4	0.88
4	Disabled Truck	14	0	2	0.71
5	Vehicle Fire	7	0	1	0.43
6	Disabled Car in Travel Lane	0	n/a		
7	Fatal Incident	0	n/a		
8	Cargo Spill	4	0	2	0.50
9	HAZMAT	3	0	1	0.33
10	Weather Related	0	n/a		
11	Const./Maint.	55	0	2	0.56

Table 6.5
Number of Fire Engines Required for All Incidents [2, 12]

Fire Engines

Number	Incident Type	Number of Incidents	Minimum	Maximum	Average
1	Road Hazard	24	0	2	0.17
2	Property Damage	438	0	2	0.14
3	Personal Injury	250	0	2	0.84
4	Disabled Truck	14	0	3	0.14
5	Vehicle Fire	7	0	2	1.14
6	Disabled Car in Travel Lane	0	n/a		
7	Fatal Incident	0	n/a		
8	Cargo Spill	4	0	3	1.50
9	HAZMAT	3	0	1	0.67
10	Weather Related	0	n/a		
11	Const./Maint.	55	0	2	0.07

Table 6.6
Number of Ambulances Required for All Incidents [2, 12]

Ambulances

Number	Incident Type	Number of Incidents	Minimum	Maximum	Average
1	Road Hazard	24	0.00	1.00	0.08
2	Property Damage	438	0.00	2.00	0.14
3	Personal Injury	250	0.00	4.00	1.17
4	Disabled Truck	14	0.00	0.00	0.00
5	Vehicle Fire	7	0.00	2.00	0.14
6	Disabled Car in Travel Lane	0	n/a		
7	Fatal Incident	0	n/a		
8	Cargo Spill	4	0.00	2.00	1.00
9	HAZMAT	3	0.00	0.00	0.00
10	Weather Related	0	n/a		
11	Const./Maint.	55	0.00	2.00	0.09

Table 6.7
Number of Arrowboards Required for All Incidents [2, 12]

Arrowboards

Number	Incident Type	Number of Incidents	Minimum	Maximum	Average
1	Road Hazard	24	0.00	2.00	0.25
2	Property Damage	438	0.00	3.00	0.04
3	Personal Injury	250	0.00	2.00	0.08
4	Disabled Truck	14	0.00	2.00	0.50
5	Vehicle Fire	7	0.00	0.00	0.00
6	Disabled Car in Travel Lane	0	n/a		
7	Fatal Incident	0	n/a		
8	Cargo Spill	4	0.00	4.00	1.50
9	HAZMAT	3	0.00	0.00	0.00
10	Weather Related	0	n/a		
11	Const./Maint.	55	0.00	1.00	0.20

Table 6.8
Number of Police Vehicles Required for All Incidents [2, 12]

Police Vehicles Number	Incident Type	Number of Incidents	Minimum	Maximum	Average
1	Road Hazard	24	0.00	3.00	0.96
2	Property Damage	438	0.00	7.00	1.22
3	Personal Injury	250	0.00	8.00	1.71
4	Disabled Truck	14	0.00	4.00	1.07
5	Vehicle Fire	7	0.00	2.00	1.29
6	Disabled Car in Travel Lane	0	n/a		
7	Fatal Incident	0	n/a		
8	Cargo Spill	4	0.00	6.00	4.00
9	HAZMAT	3	0.00	2.00	1.67
10	Weather Related	0	n/a		
11	Const./Maint.	55	0.00	3.00	0.95

Table 6.9
Average Number of Response Units for Personal Injury Incidents [2, 12]

Personal Injury Incidents No. of Injuries	No. of Incidents	Avg. No. of Cars Involved	Avg. No. of Wreckers	Avg. No. of Fire Engines	Avg. No. of Ambulances	Avg. No. of Arrowboards	Avg. No. of Police Vehicles
1	160	2.08	0.79	0.89	1.09	0.08	1.68
2	36	2.19	1.03	0.92	1.19	0	1.61
3	11	2.08	1.64	1.55	2.09	0	2.00
4	12	3.25	1.33	0.83	2.08	0	2.25
5	4	2.00	2.00	1.25	2.75	0	2.25

traffic incidents. The following resource allocation rule base has been constructed based on statistical analysis of the resources used for different types of incidents; expert knowledge that exists in the form of manuals also was used. (The few cases in which no information is available need to be studied.) An

Personal Injury Incidents

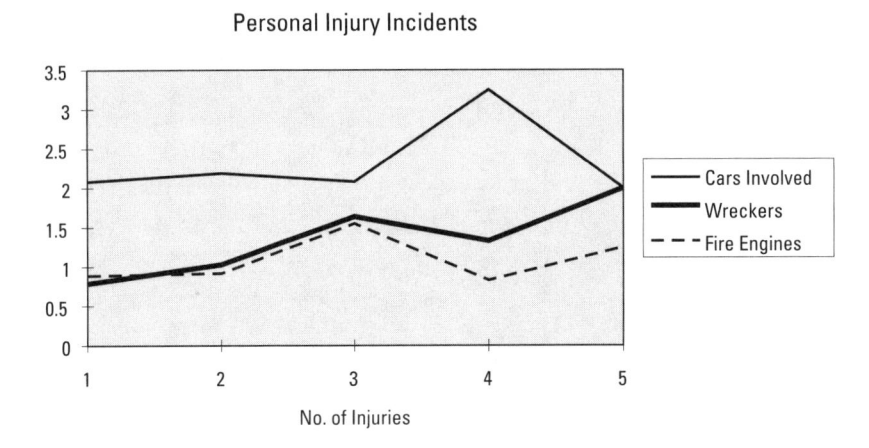

Figure 6.6 Average number of wreckers and fire engines required in personal injury incidents [2, 12].

Personal Injury Incidents

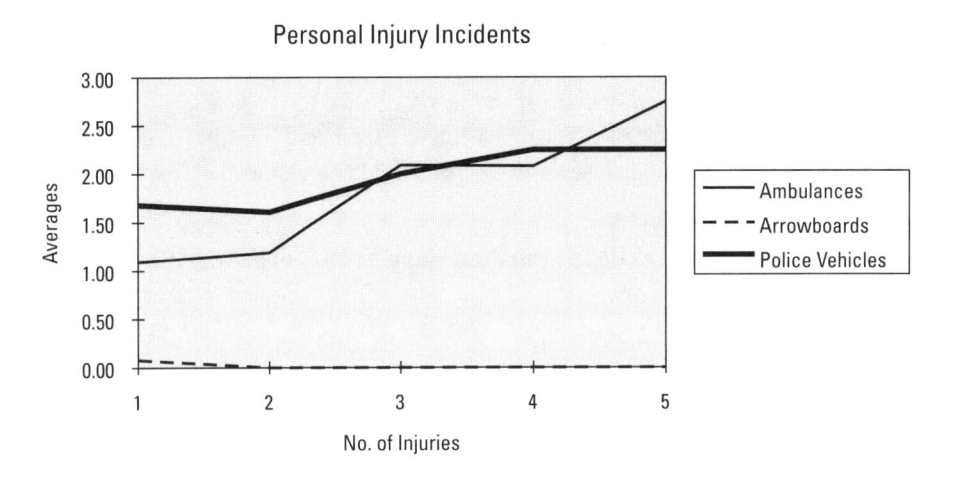

Figure 6.7 Average number of ambulances, arrowboards, and police vehicles required in personal injury incidents [2, 12].

important rule of thumb adopted by the experts is that resources dispatched should never be less than what are required, because such a situation could be disastrous in certain incidents. In development of the rule base, that has been kept in mind, and resources have always been assigned more generously (than what is obtained from the statistical analysis) in most cases. Tables 6.10 through 6.13 show the methodology employed to determine the resource

Property Damage Incidents

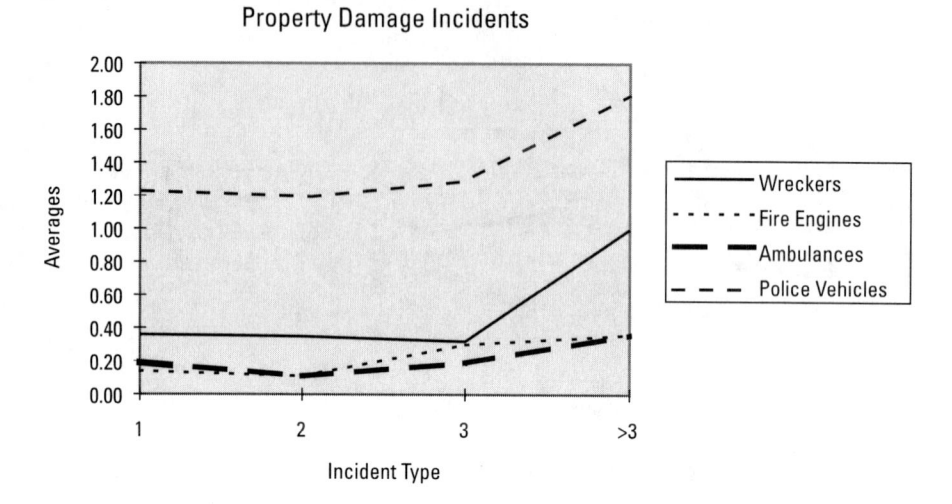

Figure 6.8 Average number of response units required for property damage incidents [2, 12].

allocation rules and whether they are based on expert knowledge or statistical analysis. For incidents other than those listed in the tables (i.e., lanes blocked, HAZMAT, vehicle damage/overturned truck, cargo spill, weather related, and disablement), statistical analysis or expert knowledge is not available.

6.4.4 Implementation of Response Rule Base as Part of WAIMSS

The response-related statistical results and the response procedures discussed in this chapter were used as the basis for development of the response rule base. The rule base constitutes the core of the WAIMSS response module. The response module of WAIMSS suggests suitable measures to clear an incident.

Table 6.10
Resource Allocation Rules for Incident Type: Fatalities [2, 12]

If	Then Send		
No. of Fatalities	Ambulances	Police Vehicles	coroner
1	1	2	medvac if
2	2	2	necessary
3 or more	3 or more	3	

Table 6.11
Resource Allocation Rules for Incident Type: Injuries [2, 12]

If	Then Send		
No. of Injuries	Ambulances	Police Vehicles	
1	1	2	medvac if
2	2	2	necessary
3	2	2	
4	3	3	
5	3	3	

Table 6.12
Resource Allocation Rules for Incident Type: Vehicle Fire [2, 12]

If		Then Send
Vehicles Involved		Fire Engines
No. of Vehicles	Type	
1	C	1
	T	2
2	C-T	3
	C-C	2
	T-T	3
3	C-C-T	3
	C-T-T	4
4 or more	—	4 or more

* from SA and EK

The module utilizes interactive queries on the availability of resources while a plan to clear the incident is determined.

WAIMSS also maintains a database of personnel to be contacted during incidents. An important challenge is to keep that database updated, since names and numbers of incident management personnel change constantly. The database, which can be easily accessed from the Response Window of WAIMSS

Table 6.13
Resource Allocation Rules for Incident Type: Property Damage [2, 12]

If	Then Send		
No. of Cars Involved	Wreckers	Police Vehicles	
1	1	2	department of
2	2	2	public works if
3	3	2	necessary
>3	3 or more	2	

* from SA

by clicking the appropriate buttons, provides important contact information to all agencies.

6.5 Summary

The case study presented in this chapter was aimed at understanding the incident response problem from a new angle. The incident response procedure was analyzed and a simple procedure for developing response plan was formulated. A statistical analysis of the response requirements for various incident types based on the incident survey conducted in northern Virginia helped to clarify the needs for resource allocation. A computer GIS-based program was then devised and implemented as part of WAIMSS.

The Arc/Info-based GIS system has online resource allocation information, agency contact information, location of response centers, and the best center choice with respect to an incident. The software also offers the capability to edit and add new centers, update attribute information, and update contact information.

The response module presented in this chapter can be further enhanced by implementing some of the following tasks:

- More response center coverages (such as towing companies) and more attribute data, such as the street address and phone number of each center, can be added to the existing coverage.
- A detailed real-time resource inventory database can be developed for each response center. That would help keep track of available resources

at each center. For example, if a fire station has four fire engines and two have been dispatched for an incident, the database would show that two fire engines are available.

Review Questions

1. Estimate the duration of four distinct components of incident response times in your area. Which one do you think is easier to reduce in the short run?

2. Draw a deterministic queuing diagram and show the four different components of the response time on the diagram. Write a computer program that estimates the delay for different response time components.

3. Develop a response procedure specific to your area. (Hint: Use the conceptual response procedure discussed in this chapter.)

4. Discuss the possible institutional problems that might occur among different agencies involved in incident management. Make a comprehensive list.

References

[1] Zografos, K. G., T. Nathanail, and P. Michalopoulus, "Analytical Framework for Minimizing Freeway-Incident Response Time," *J. Transportation Engineering*, ASCE, 1993.

[2] Dhingra, N., *Incident Response*, M.S. thesis, Virginia Polytechnique and State University, 1996.

[3] Virginia Department of Transportation (VDOT), *Northern Virginia Freeway Management Team Operating Manual—A Regional Plan for Traffic Management on Northern Virginia Freeways*, April 1990.

[4] Stack, S. G., and G. R. Ritchie, *Caltrans District 12 Real-Time Expert System for Freeway Incident Management*, Final Report, Univ. of Calif., Irvine, 1993.

[5] I-95 Corridor Coalition, *Incident Management, Detection, Response and Operations*, Regional/Corridor Coalition Agencies, Final Working Paper, I-95 cc 2-94-03, 1995.

[6] I-95 Corridor Coalition, *Incident Management, Detection, Response and Operations*, State-of-the-Practice Reports for the I-95 Corridor Coalition Agencies, Final Working Paper, I-95 cc 2-94-03, 1994.

[7] Zografos, K. G., C. Doulegeris, and L. Chaoxi, "Simulation Model for Performance of Emergency Response Fleets," *Transportation Research Record* 1452, pp. 27–34.

[8] Nathanail, T., "Evaluation Platform for Integrated Incident and Traffic Management Systems," *Proc. IFAC Transportation Systems Conf.*, Chania, Greece, 1997.

[9] Zografos, K. G., and G. M. Vasilakis, "Theoretical and Practical Issues in Incident Response Operations," *Proc. IFAC Transportation Systems Conf.*, Chania, Greece, 1997.

[10] Eaton, D. J., et al., "Determining Emergency Management Services Vehicle Deployment in Austin, Texas," *Interfaces* 15, 1985, pp. 96–107.

[11] *Incident Management News*, SIM Committee Publication, 1996.

[12] Kachroo, P., K. Ozbay, Y. Zhang, and W. Wei., "Development of a Wide-Area Incident Management Expert System," (work order #DTFH71-DP86-VA-20) Final Report, FHWA, 1997.

7

Traffic Diversion for Real-Time Traffic Management During Incidents

7.1 A Scenario

Consider the following scenario for a small portion of Fairfax county in the northern Virginia network shown in Figure 7.1. The network covers 9.5 miles of the I-66 freeway section, and the neighboring arterials consists of 57 nodes and 111 links. A major accident involving multiple cars and trucks occurs on I-66 during a peak hour [1]. One of the two lanes and the shoulder of I-66 are closed due to the incident. Although detection and verification of the incident take no more than 10 minutes and response begins immediately, restoration of the interstate traffic to normal flow takes at least 1 hour. Cars begin to build up from the point where the accident occurred. Drivers approaching the scene feel the effects of the incident as traffic slows. The VMSs display "Major Accident Ahead" and "Use Alternate Route." The traffic management center at Arlington must quickly make the decision to divert cars heading toward the incident location. The traffic management center is responsible for coordinating the response as well as for the selection and best implementation of the traffic management measures.

7.2 The Solution Approach

Most accidents are minor and do not require drastic measures such as diversion. However, for a major accident similar to the one described in the scenario,

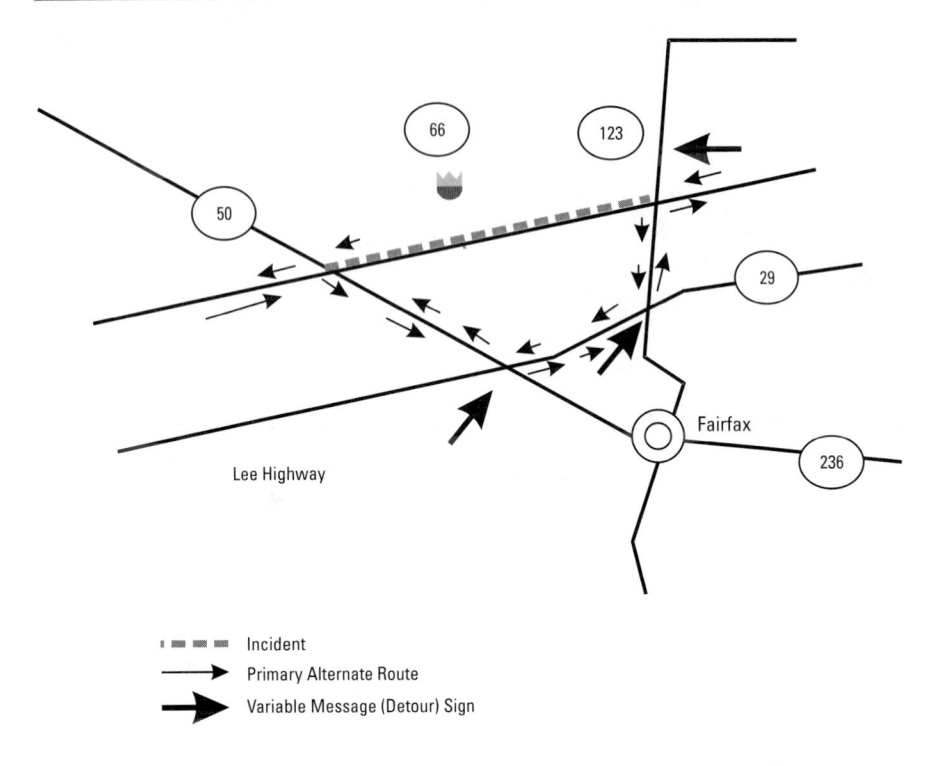

166   Incident Management in Intelligent Transportation Systems

Figure 7.1 Sample Fairfax network for scenario [2].

diversion often is the only way to alleviate congestion. The solution approach to such a situation calls for the development of an integrated, multilayered framework to assist traffic management operators in developing and implementing diversion strategies in real time. The traffic manager at the TMC needs to perform these steps quickly.

1. First, accurately predict the duration of the incident and decide if there is a need for diversion.
2. Then decide the diversion point(s) and alternative routes to be used for diversion. Given time and equipment limitations, it generally is difficult to implement diversion for more than one diversion point. That is why it is extremely important to choose the best diversion point(s) for the specific incident and determine the best routes for the chosen diversion point(s).
3. Finally, actuate the VMSs in a way that allows the traffic flow on the alternative routes to be controlled dynamically and avoid overreaction of the motorists to the incident and routing advisory.

The problem may be made more complex by a decision to change signal timing plans on the arterials of the selected diversion routes in real time. That type of change on the arterials to accommodate additional diverted traffic definitely reduces congestion on the arterials. However, that would also require advanced real-time signal optimization algorithms and appropriate infrastructure. Although several cities in the United States have installed advanced signalization hardware, most of the real-time signal optimization algorithms are still under development and not widely deployed. The lack of well-tested real-time signal optimization algorithms makes it impossible to use the advanced communication and signalization infrastructure to its full potential. Federally funded projects such as Real-Time Adaptive Control Strategies (RT-TRACS) [3] currently are working on the development of advanced online signal optimization systems. RT-TRACS is still being developed and is not yet operational.

Several research and development projects are attempting to develop operational diversion strategies for real-life situations. Among them are the dynamic traffic assignment project known as "development of an operational dynamic traffic assignment system" (Mahmassani, 1994) and a VDOT project titled "development of traffic control strategies for the Suffolk traffic management system." The first project attempts to build a comprehensive dynamic traffic assignment system to address many situations, including traffic diversion during incidents, traffic routing under normal traffic conditions, and planning purposes. The second project attempts to develop traffic control strategies for a two-alternative-route system using the results of offline traffic simulation studies. However, as yet, none of the new projects have proposed a practical and working framework to deal with real-time diversion problem during freeway incidents. The two projects mentioned here, as well as several other projects, are still being researched and developed and cannot be implemented in a real-world environment in the immediate future.

The discussion of a network-wide optimization scheme for dealing with the diversion problem is beyond the scope of this book. This chapter simply presents some of the more practical studies and projects that deal with the diversion problem during incidents. It also presents the expert system approach adopted by WAIMSS to determine possible diversion routes in real time. Our goal is to provide useful information for developing diversion strategies during major incidents. However, we believe that with the proliferation of advanced real-time routing and signalization algorithms coupled with the required infrastructure, more sophisticated networkwide diversion strategies than the ones discussed here will be implemented in the future.

Section 7.3 describes some of the most relevant research projects dealing with the real-time diversion problem and its solutions.

7.3 Traffic Diversion

Real-time diversion and routing of traffic is one of the most efficient ways of relieving non-recurrent traffic congestion. Several models have been developed for determining diversion routes and diverting traffic onto those routes. Expert systems, feedback control, and mathematical programming models are among the approaches that have been used for developing real-time diversion and routing strategies.

A careful review of literature reveals several models of different complexity used to determine diversion routes for real-time applications. One of the first comprehensive studies on diversion is conducted by Hall [4]. That study explores the feasibility of real-time diversion in an intercity corridor between Baltimore and Washington, D.C. Hall uses simulation to evaluate several diversion strategies for different scenarios of various states of the system. He proposes the use of either simulation of corridor traffic (SCOT) or the New Jersey Turnpike model developed by Sperry Rand Corporation for analyzing the different diversion scenarios. Among the system features Hall [4] simulates are:

- Traffic surveillance system, including traffic detectors, aerial surveillance, CCTV, call boxes and telephones, and highway patrols;
- Communication system that allows timely exchange of traffic and diversion data between the accident site and the traffic control center;
- VMSs to disseminate traveler information such as congestion length and cause and alternative route information to the drivers.;
- Traffic control center to act as the command control center to manage traffic after the incident occurs.;
- Different incident scenarios such as closure of the Baltimore/Washington Parkway at different points of the parkway for different levels of demand.

As a result of his study, Hall concludes that real-time diversions can provide a more comfortable driving environment and minimize the consequences of incidents. He also points out some perceived drawbacks, such as the difficulty of developing diversion plans in real time and the cost of implementing those plans.

Another diversion model [5] is also specific to the Baltimore beltway. That model uses Sperry-developed event-scanning traffic simulation based on the hydrodynamic traffic model with the capability of simulating multiroadway freeway systems. It proposes a simple control mechanism that tries to minimize delay difference or delay rate difference at alternative routes by switching the diversion messages on and off. The system suffers from the network-related weakness in that, in the event of an incident of sufficient severity on any link of the beltway, only one alternative route can be used for the entire beltway. Furthermore, the proposed real-time traffic routing controller is heuristic in nature, and there is no guarantee that it will work under congested conditions.

Another study [6] describes a traffic management plan at Los Angeles aimed at bringing about voluntary traffic diversion upstream of a section of freeway that needed to be closed for 6 hours for maintenance operations. A diversion plan specifically prepared for that closure has actually been implemented. The authors report that the operation has been carried out smoothly, with no major problems. Although this specific example shows that preplanning works for managing major special events that are known before they occur, it does not address the issue of generating diversion plans in real time.

In a more theoretical study, Ketselidou presents an expert system model for postincident traffic control [7]. The aim of the study is to demonstrate the working of an expert system–based diversion system on a test network with desirable attributes such as the availability of alternative routes. The model developed is based on pre-determined weights assigned to links depending on the time of the day and the historical traffic volumes. Points at which diversion can be initiated and the potential destinations for particular links also are determined beforehand. The search process then carries out an exhaustive search based on those preset thresholds. The best diversion strategy is chosen to be the one that balances the unused capacity according to three rules listed in order of decreasing priority:

1. Wherever possible, the link demand will be kept below the capacity for all links in the system.

2. Whenever the unused capacity is less than a certain threshold, control will be exercised to balance the unused capacities on an adjacent pair of links provided rule 1 is not violated.

3. Diversion will be terminated when neither rule 1 nor rule 2 is violated.

Other criteria are employed for selection of alternative routes include proximity to main corridor routes, usefulness for access to ultimate destination, driving quality on route, impact on adjoining land use, and jurisdictional problems.

This study uses the Long Island, New York, network, which is basically a freeway corridor which has great potential for alternative-route assessment because of the availability of a number of freeways and associated arterials as the test network. The system is evaluated and tested by comparing the results obtained from the developed expert system and a macroscopic traffic simulation package called TRAFLO developed by the FHWA. The results of the evaluation and tests are reported to be fairly consistent with the results of the simulation runs.

Although the explicit consideration of factors other than traffic volumes in the determination of diversion routes is its salient feature, the model is limited by several deficiencies, such as the preselection process used to determine alternative routes. The pre-selection process is similar to the approach adopted by several DOTs, including the VDOT. However, pre-selected routes are not always the best routes for diversion under real network conditions. Unexpected congestion on a preselected route can create more problems for the diverted and the current traffic than the problems created due to the actual accident on another route. Thus, pre-selection of routes sometimes can be the weakest link of the whole diversion process and seriously undermine the effectiveness of traffic diversion.

The model developed by [8] considers information on traffic volumes in the arterials as an important influence in determining the diversion routes. Unlike the model [5], their model is dynamic in the sense that the diversion routes are updated in time. Another appealing characteristic is that trends in traffic conditions (like anticipation of the afternoon peak) are taken into account by applying historical patterns to real-time data to produce forecast volumes. The functioning of the algorithm for dynamic network diversion is based entirely on a pre-defined set of diversion routes identified for each link of the freeway. The algorithm focuses on checking each route's volume-capacity ratio (v/c). If v/c is less than a predefined critical value and if the time to reach the origin node of the diversion is less than the incident impact time, the route is chosen as a diversion route; otherwise, it is not considered for diversion.

While features like recognition of the dynamics of route costs, the severity of incidents and their impact on diversion route selection, and the use of historical information for prediction make this model a significant improvement over the others, it does have its share of deficiencies. The three major deficiencies associated with this modeling process are the following:

- The dynamics of diversion factors, such as maintenance, bad weather, and jurisdictional issues other than traffic volumes, can render some of these preselected routes infeasible for real-time traffic management.

- Some other diversion routes that are not included in the set of predetermined routes may often be more suitable for diversion than the preselected routes based on the real-time traffic conditions.
- In the case of real networks with hundreds of freeway links, explicit enumeration of alternative routes for each freeway link may be a laborious, inefficient, and unrealistic task. Instead, efficient time-dependent disjoint or partially disjoint route determination algorithms can be used to generate alternative routes in real time.

Furthermore, this model will not be responsive to sudden changes in the traffic flow patterns, because it uses only the historical data for analysis and determination of diversion rates. Certain safety considerations or occurrence of secondary incidents can seriously affect the validity of the solution provided by this model.

User compliance is another important issue with respect to diversion. Although the routes selected on which to divert drivers may be very good, the diversion process may fail if the confidence level of the drivers is not high. Taylor [9] indicates some of the salient features that can be used to generate diversion routes using a knowledge-based expert system. The problem data is divided into two parts:

- Basic trip data about the origin, destination, and departure time;
- Information about traveler characteristics that may be needed.

The assumption is that in a road network with many alternatives available, the best decision (the preferred route) largely depends on the attitudes and preferred behavior patterns of drivers. Because different drivers have different characteristics, the diversion information should be generated for different types of drivers. For example, a driver unfamiliar with the area might want to drive only on the major freeways and arterials, unlike an experienced driver, who may be ready to use a route that passes through the local street system. Depending on the trip purpose and the type of the driver, sometimes a route that offers a guaranteed travel time savings over other routes may be more appealing. For example, most truck drivers dislike slowing down at intersections or getting stuck in local traffic. They prefer to travel on a route that has free-flow traffic conditions even if it takes longer. In another route-choice study [10], the authors develop a statistical route choice model using the data from a road network in Torino, Italy. The results of that analysis show that the significant variables in an individual's route choice model are "the travel times on primary roads and on secondary roads," total path length, number of left turns, and signalized intersections on the selected path. Those factors can be thought of as the factors that determine the disutility of each alternative route. Drivers will select the path with the lowest disutility.

It is clear that the success of an efficient traveler information system depends entirely on its ability to provide satisfactory advice to travelers. That, in turn, requires a good understanding of users' preferences vis-à-vis route diversion decisions. In a recent study at Texas A&M [11], it was determined that for the results of traffic diversion to be predictable, it is essential to model individual users' preferences and corresponding thresholds. That study concludes that drivers base their time-saved thresholds on how much they disliked one of the recommended routes and not what they preferred. The results indicate that transportation agencies will continue to find it difficult to predict how motorists will respond to specific travel time information unless individual preferences and diversion thresholds can be obtained directly from motorists.

A related study [12] reports that drivers who listened to radio traffic reports have a "greater propensity to switch routes." Drivers' diversion propensity was also studied extensively by [13]. The stated preference approach is used to study drivers' diversion propensity. The empirical model based on a survey of downtown Chicago automobiles indicates the relative importance of each variable in determining the diversion route. The results of the model show that drivers are more willing to divert during incident conditions rather than for recurrent congestion. The second result is that commuters are more willing to divert to an alternative route that is familiar, passes through safe neighborhoods, and has no traffic stops. Furthermore, drivers of higher income are found to be more willing to divert, possibly due to their higher value of time.

7.4 Diversion System Architecture of WAIMSS

Based on the practical needs of traffic operation centers for performing real-time traffic diversion, a diversion system architecture has been developed and implemented as part of WAIMSS [14]. This section describes in detail the development and implementation efforts of a practical and real-time dynamic traffic assignment/routing system that will be used by a traffic control center during incident conditions. To achieve that goal, the system must:

- Be realistically and robustly designed to tackle real world problems and to model the real diversion process during incidents;

- Be fast enough to develop effective traffic diversion strategies in real-time;
- Make use of both the historic and real-time sensor data;
- Be flexible so it can be used in different cities and towns for different types of incidents.

To develop a realistic and effective system framework for traffic diversion during incidents, first we have to determine the different steps followed during the diversion process:

1. After an incident has been detected and verified, predict the duration of the incident based on the incident characteristics.
2. Based on the prediction of incident duration, predict the length of the delay that will be caused by the incident.
3. If the delay is more than a threshold value, initiate the diversion based on the diversion plans developed either offline or online.
4. Implement and control the diversion using different tools, such as VMSs and HAR.

Each step of the diversion process requires realistic and efficient models and algorithms for real-time implementation. The rest of this chapter discusses and describes in detail the models and algorithms that have been developed as part of the WAIMSS research work.

7.4.1 System Components

The diversion system architecture is divided into the three basic modules: the diversion initiation module, the diversion strategy planning module (heuristic network generator), and the diversion control/routing module (Figure 7.2). Each module has several submodules. The diversion initiation module has two submodules: the incident duration prediction sub-module and the incident delay estimation sub-module. The diversion strategy planning module, also called a heuristic network generator [15-16], has link elimination, route generation, and route prioritization submodules. The diversion control/routing module has several submodules in the form feedback control models that use real-time sensor data to control the diversion process to achieve a preset diversion objective.

```
┌─────────────────┐
│ 1               │
│ Diversion initiation │
│ module          │
└────────┬────────┘
         │
         ▼
┌─────────────────┐
│ 2               │
│ Network generator │
└────────┬────────┘
         │
         ▼
┌─────────────────┐
│ 3               │
│ Diversion control/ │
│ routing module  │
└─────────────────┘
```

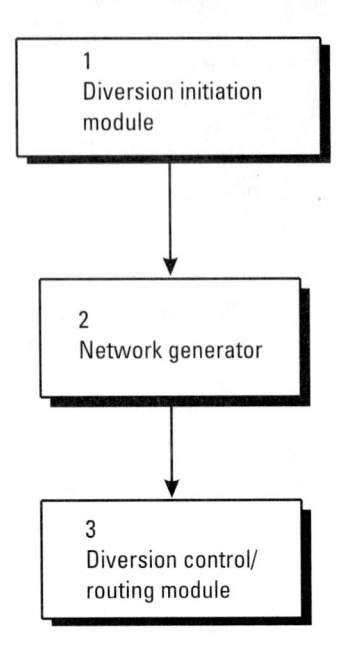

Figure 7.2 The three modules of the proposed framework for the diversion process during nonrecurring congestion.

7.4.2 Diversion Initiation Module

Traffic diversion is an effective tool to alleviate non-recurrent congestion, but the decision to use it for real-world application is not a trivial one. In this research, we propose the diversion initiation logic depicted in Figure 7.3. According to this logic, the incident prediction module first predicts the duration of the actual incident. The models used to perform the incident duration predictions are developed using real incident data collected by the incident management personnel from northern Virginia [14, 17]. (Model development efforts and the validity of the models are discussed in detail in Chapter 4.) The incident predictions are then used to predict delays on the freeway. Mastbrook proposed a complete general framework for delay prediction [18]. That framework, which requires the use of macroscopic traffic simulation, has not been fully implemented as part of WAIMSS due to the high computational costs of traffic simulation. An extended version of a more simplistic approach using the deterministic queuing model for delay estimation [19] may be adopted as an alternative to traffic simulation. (That approach also is discussed in Chapter 4.) The delay is predicted using the predicted incident duration as well as the lane blockage information and the traffic demand during the time it takes to clear the incident.

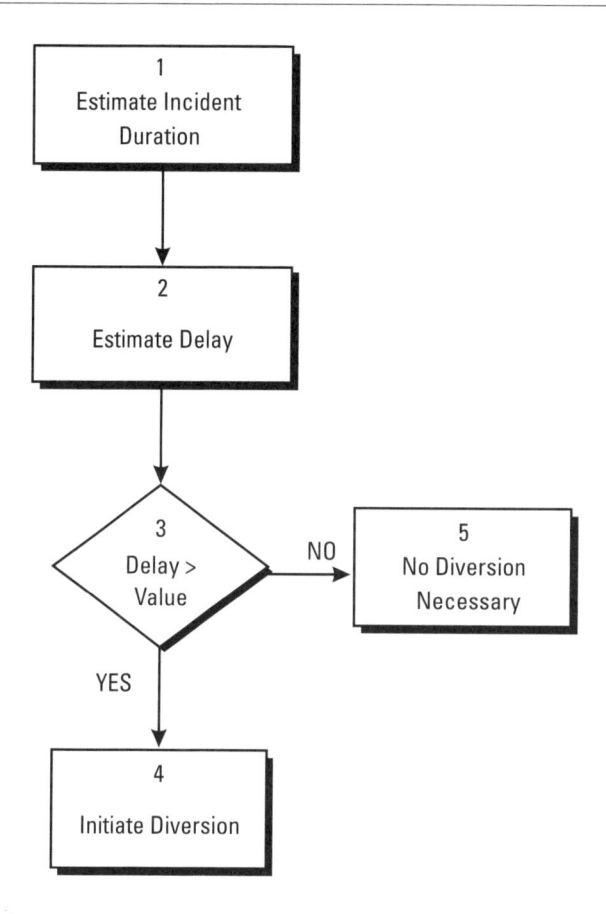

```
         ┌─────────────────┐
         │        1        │
         │ Estimate Incident│
         │    Duration     │
         └────────┬────────┘
                  ▼
         ┌─────────────────┐
         │        2        │
         │  Estimate Delay │
         └────────┬────────┘
                  ▼
              ╱   3   ╲         ┌─────────────────┐
             ╱  Delay > ╲  NO   │        5        │
             ╲  Value   ╱──────▶│   No Diversion  │
              ╲       ╱         │    Necessary    │
                │YES            └─────────────────┘
                ▼
         ┌─────────────────┐
         │        4        │
         │ Initiate Diversion│
         └─────────────────┘
```

Figure 7.3 Logic of the diversion initiation module.

The severity of delay is used as the mechanism to trigger the diversion process. The threshold values are set up using the expert knowledge obtained from the northern Virginia incident management personnel. However, the threshold values will be different for different networks, depending on the level of daily recurring congestion, the incident management procedures of the area, and other site-specific considerations.

7.4.3 Diversion Strategy Planning Module (Heuristic Network Generator)

It is clear that under incident conditions, it is not always necessary to consider the entire network in generating effective diversion strategies. An effective approach in addressing that problem is to determine the incident impact zone, that is, the portion of the overall network affected by the specific incident. The incident impact zone will be smaller than the entire network for the majority of incidents. Also, it is well known that some of the links are not suitable for diversion for many reasons, including the already high v/c ratios, ongoing construction, lack of equipment needed to route motorists, and jurisdictional considerations

The network generator is a knowledge-based expert system developed to address those issues by preprocessing the network and determining a set of feasible routes that can be used for diversion (Figure 7.4). The knowledge base is developed using information obtained from experts in northern Virginia and the existing literature. The knowledge base consists of simple rules that determine the incident impact zone based on the incident delay and incident

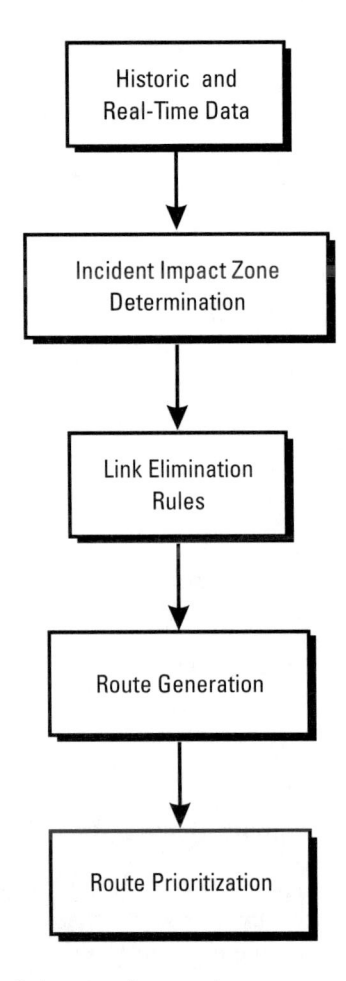

Figure 7.4 Logic of the heuristic network generator.

characteristics and rules that eliminate links that are not feasible for diversion based on the real-time and historic data and the network-specific knowledge. The hybrid architecture of WAIMSS, which consists of a GIS, Arc/Info, an expert system shell, Nexpert-Object, and C programs, is also used for the implementation of the network generator. That architecture supports the real-time working of the system by allowing the real-time manipulation of large network data through the use of GIS and supporting the decision-making process through the use of the expert system. The architecture also allows a client/server implementation of the network generator, which proves to be useful in the multiagency environment of the incident management process.

7.4.4 Diversion Control/Routing Module

The heart of the proposed diversion system is the effective and real-time control of traffic. The major goal of the diversion control/routing module is to control diverted traffic volumes to prevent overloading of some alternative routes and to reach a predetermined goal, such as system optimal or user equilibrium. Unlike other similar models that attempt to control traffic using off-line optimization approaches, the diversion module of WAIMSS achieves the goal of real-time traffic control by employing feedback control. The feedback control models developed in this research still are applied only to simple test networks and need to be enhanced and calibrated for real-world implementations. On the other hand, they present a new and efficient alternative to the existing networkwide optimization models. These control models will use the sensor data to decide about the control commands given to motorists. The control commands will be disseminated through VMSs or HAR. The control commands will be revised at every sampling time based on the sensor data obtained from the traffic sensors on the roadway.

The feedback control models developed in this research are specifically designed for controlling the point diversion of traffic to multiple alternative routes determined by the heuristic network generator. The system dynamics and modeling issues of the feedback control models are described in Chapter 8.

Section 7.5 describes in detail the functions and theory of the network generator.

7.5 Functions and Theory of the Network Generator

The network generator has several distinct functions, including incident impact zone determination, link elimination, route generation, and route prioritization functions. The incident impact zone determination and link elimination functions have their roots in the earlier work done by transportation researchers to reduce the computational complexity of network flow problems [20-22]. Determination of the incident impact zone and link elimination aim to reduce the size of the network by focusing on a smaller portion of the network and by eliminating infeasible links in the network. Thus, one may broadly say that the network generator also functions as a type of network aggregation tool. In general, however, the network generator acts as a flexible, real-time decision support system for supporting the overall real-time diversion process and determining alternative routes for diversion. That goal is achieved by:

- Using heuristic rules to determine the incident impact zone and to eliminate infeasible links;
- Employing an expert system shell, Nexpert-Object, which allows efficient coding of the heuristic rules;
- Using a GIS, Arc/Info, to manipulate and process large amounts of network data in real time that otherwise are impossible to handle in real time;
- Employing state-of-the-art data and command-level software bridges between Arc/Info and Nexpert-Object that are developed as part of overall WAIMSS software package, thus allowing the real-time data and command exchange that is vital for the dynamic alternative route generation;
- Supporting real-time communication of involved agencies through the client/server architecture used for its implementation;
- Being an integral part of the implementation architecture of the WAIMSS software package and providing a unifying framework for other components of WAIMSS.

The software implementation of the network generator is an important portion of the overall concept of the network generator as well as the WAIMSS software package.

7.6 Network Aggregation Models

Several network aggregation models have been developed with the intention of reducing the prohibitively large size of transportation networks that will be used for transportation planning studies. Understanding those earlier concepts are necessary to introduce the link elimination concept of the network generator module. Network aggregation can be defined as the task of condensing a given network into a smaller one that can be managed efficiently as well as preserving the desired characteristics of the original network. There are two main approaches to the problem:

- *Network element extraction (NEE)* is the process of removing from the network elements identified as being insignificant according to some prespecified criterion. This method has the disadvantage of causing network disconnection.

- *Network element abstraction (NEA)* collapses the insignificant network elements into pseudo- or dummy elements.

One NEE model is based on the distribution of traffic on the network after an assignment process [22]. The links that do not carry a significant amount of traffic at the end of the assignment are identified and extracted from the network, and the performance of the remaining portion of the network is then studied. A link is deemed to be insignificant if it carries an equilibrium flow below a fraction of the maximum equilibrium flow in the network. Some of the disadvantages of this model are [22]:

- The traffic assignment process can be adversely affected by the increasing number of extracted number of links, because that may produce a set of smaller disconnected networks.

- It is clear that a possible increase of the number of origins and destinations also may increase computation complexity of the algorithm. However, this possibility of the increase of the Origin-Destination (OD) pairs due to the NEE model is found to be significant in this paper.

- The models employ a static assignment, instead of a dynamic one, to determine link volumes. Static traffic assignment models can be used for long-term planning purposes, but their applicability for dynamic conditions is questionable.

- Further real-time dynamic traffic conditions such as incidents or bad weather may require factors other than merely *v/c* ratios as the governing criterion for performing an effective network aggregation process, especially for traffic diversion during incidents.

- The NEE model is incapable of handling real-time conditions of traffic, because it was developed mainly for use in project-planning applications in which different scenarios can be studied without significant computational effort.

NEA methodology is presented in a paper by Eash et al. [21] for the northern Illinois network, which also compares regional planning to the sketch planning approach. The characteristics used for the comparison are the total vehicle-miles, vehicle-hours, and average speeds. One of the obvious impacts of the abstraction was the immediate increase in the number of intrazonal trips for the sketch planning case because the zones became bigger as a result of aggregation process. A methodology is described to account for that increase, in which the vehicle-miles for each of the two cases are studied and the results of the regional assignment are adjusted according to the difference. That provides a basis for comparing the results of the two techniques. The adjusted values are presented in Table 7.1. The additional intrazonal trips are assigned to the same minimum time paths, in the same proportions used in the regional traffic assignment. Some of the general conclusions that can be drawn are:

- Different intrazonal trips in the two assignments did not significantly affect the results.
- Computational performance of the traffic assignment is more seriously affected by the coding of the arterial network.

Table 7.1
Sketch Planning Versus Regional Assignment (*From:* [21], with permission from the author.)

No.	Item	Sketch Planning Assignment	Regional Assignment
1)	Network Nodes	820	12040
2)	Network Links (One-way)	2422	37065
3)	Computing Time (CPU)	3 min. 45 sec.	163 min. 7 sec.
4)	Freeway Average Speed	33.1 mph	36.6 mph
5)	Arterial Average Speed	26.8 mph	24.5 mph
6)a	Veh.-hours: Freeway	104,962 (29%)	81,446 (22%)
6)b	Arterial	261,048 (71%)	286,664 (78%)
7)a	Veh.-miles: Freeway	3,475,759 (33%)	2,981,913 (30%)
7)b	Arterial	7,000,609 (67%)	7,025,788 (70%)

- The overall results, however, compared quite well. Therefore, sketch planning assignment is found to be adequate for estimating most highway travel characteristics, including operating costs, emissions, and gasoline consumption.

The methodology used for aggregating the network itself seems heavily dependent on the individual network characteristics. However, the performance results of the NEA algorithm presented in [21] has to be studied for other areas of the country before general conclusions can be drawn with any certainty.

In a similar study carried out by Bovy and Jansen [20], the network is aggregated and the effects on the car traffic assignment module are empirically investigated. Three network models were developed for the road network of Eindhoven: a fine, a medium, and a coarse model. They also present results of the all-or-nothing and equilibrium assignments based on the sensitivity of link load estimates to the different cases. The important conclusions drawn from that research are:

- Extreme reductions in network size can lead to significant errors in the estimations, while a medium-level network consisting of all arterials and collectors appears to give results that can hardly be improved.
- Even in only slightly congested networks, an equilibrium analysis seems to give better results than the all-or-nothing model.

This brief review of the network aggregation methods shows that although they represent an important step for the conceptualization of the network generator module, they are not suitable for real-time applications. An important difference between the network aggregation models presented here and the network generator is that aggregation models emphasize the planner's perspective, whereas the model under development emphasizes the real-time characteristics. That is reflected in the fact that the aggregation models emphasize the importance of retaining the original characteristics of the network. That is an important characteristic because for a planner any new modifications, such as new construction, always occur on the original network. Saving computation time is only a secondary consideration for these studies. For diversion, one is interested in developing a set of routes for a short period (in most cases) in real time. Therefore, in addition of considering v/c ratios, other factors derived from expert knowledge are incorporated into the model.

Moreover, the review of some of the state of the practice of real-time diversion techniques used reveals that currently existing decision support merely selects diversion routes from a predetermined set of routes that are manually input into the system. The systems do not account for dynamic network conditions that may cause the predetermined set of routes not to be feasible. In such a case, new diversion routes need to be generated in real time. The network generator is designed to achieve that using a hybrid system that consists of a GIS and an expert system that identifies such routes in real time by extracting feasible links from the original network. An additional advantage of this hybrid system is the ability to develop diversion plans offline for a variety of potential incident cases. The network generator also facilitates the editing, storing, and retrieval of these plans.

7.7 Theoretical Modeling of the Network Generator

This section discusses in detail the theoretical modeling of the network generator. For convenience in presentation style, this section begins by defining diversion strategies. That is followed by a discussion of the modules that constitute the network generator.

7.7.1 Elements and Types of Diversion Strategies

A diversion strategy can be completely described with a set of traffic flow variables in conjunction with a set of implementation tools that disseminate the strategy to the drivers. The two basic elements of a diversion strategy are the traffic flow decision variables and the traffic management tools.

The traffic flow decision variables include the following:

- Diversion volume(s);
- Starting point(s) for diversion;
- Termination point(s) for diversion;
- Diversion route(s).

The traffic management tools include:

- VMS;
- HAR;
- Detour sign(s);
- In-vehicle advisory messages.

Different values for the different traffic flow variables lead us to identify different sets of diversion strategies. The diversion strategies can be classified into the following sets:

- *No-diversion.* The no-diversion base case is a strategy that may be appropriate when the incident delay is not significant or when the alternative routes themselves are very congested.

- *Point diversion.* Point diversion is a strategy that diverts the traffic from the exit ramp immediately preceding the incident link to the entry ramp immediately following the incident link. In cases when that strategy may not be the most appropriate, the corridorwide diversion strategy process may be adopted.

- *Corridor-wide diversion.* The corridor-wide strategy attempts to divert motorists from a few exits upstream of the incident to a few exits downstream of the incident.

- *Inter-freeway diversion.* In the event of a major incident that causes several freeway links to be closed for long periods of time, an inter-freeway diversion strategy to divert traffic from one freeway to another may be appropriate.

The network generator considers all four options. The diversion strategy thus requires specification of the diversion volume(s), origin(s), and destination(s) and the best route(s) to carry the diversion volume. It is important to note that the best strategy may turn out to be the no-diversion strategy.

7.8 Estimation of Incident Impact Area

An important feature of the network generator is the incorporation of a concept called the incident impact area. The prototype expert system developed by [15] searches for alternative diversion routes within the entire network, regardless of the severity of the incident. That task is computationally expensive and is unnecessary when the incident is not severe, that is, when diversion routes around the incident area can be found. The enhancement carried out as part of the research work in [14] elaborates on that idea to define an area around the incident called the incident impact area.

The incident impact area is the area around the incident that will form the search space to find alternative routes. For an incident of high severity, the most beneficial diversion strategy may call for starting and termination points of diversion located a long distance from the actual site of the incident. For such a case, the incident impact area needs to be large enough to be able to permit the high volumes of traffic flow from the starting point(s) to the termination point(s) of the diversion. For an incident of lesser severity, it may suffice to have a relatively smaller search area for the diversion strategy. Four factors influence the size of the study area:

- *Incident severity.* The higher the severity of an incident, the larger the size of the search area needs to be.
- *Spatial distribution of network.* In a location where the road network is sparse, an incident of a given delay may need a larger area for diversion.
- *Traffic congestion levels.* If an incident of some severity occurs on a freeway link near a highly congested area, it is likely that a greater search space for diversion is needed than if it had occurred in a location around which the traffic congestion is not very high.
- *Motorist behavior.* Studies show that motorists may not be willing to divert to routes far from their original routes. Hence, motorist behavior has to be considered in determining a search area.

7.8.1 Representation of Incident Impact Area Knowledge

While expert systems can represent knowledge in many ways, including rule frames and semantic networks, the knowledge for determining incident impact areas is represented using production rules. Production rules are simple statements of the form "*If* conditions are true/ *Then* consequences are true." The production rules developed for the impact area module determination ideally should account for all four factors. Those heuristic rules use the incident severity to estimate an incident impact area. The set also includes heuristics relating to traffic congestion levels around the incident area that suggest different impact areas for peak and nonpeak conditions. This first set of rules, called "Incident Impact Area—General Rules," accounts for the first two factors. The site-specific rules that account for the next two factors are called "Incident Impact Area—Location Specific Rules."

First, an incident severity index for each incident is defined based on interviews with the experts and previous studies. The severity index is defined as a function of three important factors, namely incident duration, delay, and type, that are also used by the incident initiation logic. The concept of incident severity index is summarized in Table 7.2.

Two axes are considered to determine the impact area: the major axis and the minor axis. The major axis is defined as the direction that carries the majority of traffic that will pass through the incident site. For example, if the incident occurs on an east-west freeway, the major axis will be in that direction. The north-south direction will then be chosen as the minor axis. For the general rules governing incident impact area, the mapping of the expected delay to the major and minor axes of the incident impact zone is obtained from recent incident management–related literature. An example set of these rules used for the demonstration system is summarized in Table 7.3. The threshold values in the rules are location dependent and will vary from one location to another. Thus, it is extremely important to calibrate the threshold values in consultation with the incident management teams in the specific implementation area.

7.8.1.1 Location-Specific Rules

Location-specific impact area rules should also be based on consultations with the experts since they are mainly determined based on the network characteristics and motorist behavior in the study area. It should be noted that the determination of location-specific rules is one of the more difficult tasks because for different areas different sets of rules might be needed. The version of the network generator developed in this project does not have such location-specific rules, mainly due to the lack of expertise in determining location-specific rules. It is clear that local rules have the potential of improving the overall impact zone determination by introducing network-specific considerations into the process.

Table 7.2
Incident Severity Index Rules [14]

Incident Severity Index	Rules
Severity Index = 1	If incident_clearance_time $\leq$ 30 mins and If average_incident_delay $\leq$ 15 min/veh and If incident_type = Pre-specified_type
Severity Index = 2	If incident_clearance_time $\leq$ 60 mins and If 15 average_incident_delay $\leq$ 30 min/veh and If incident_type = Pre-specified_type
Severity Index = 3	If incident_clearance_time $\leq$ 120 mins and If 30 min/veh $\leq$ average_incident_delay $\leq$ 45 min/veh and If incident_type = Pre-specified_type
Severity Index = 4	If incident_clearance_time $\geq$ 120 mins and If average_incident_delay $\geq$ 45 min/veh and If incident_type = Pre-specified_ type

Table 7.3
Incident Impact Area: Examples of General Rules [14]

Severity Index	Impact Area Major Axis	Impact Area Minor Axis	Diversion Strategy
1			No Div
2	≤5 miles (peak) ≤4 miles (non-peak)	≤4 miles (peak) ≤3 miles (nonpeak)	Point Div / Corridor Div
3	≤8 miles (peak) ≤6 miles (non-peak)	≤6 miles (peak) ≤4 miles (nonpeak)	Point Div / Corridor Div
4	≤18 miles (peak) ≤15 miles (non-peak)	≤9 miles (peak) ≤6 miles (peak)	Corridor Div / Inter-Freeway Div

7.8.1.2 Determining the Starting and Termination Points of the Diversion

The network generator identifies the starting and termination points for the diversion based on the expected duration, delay, and associated backup due to the incident (Table 7.3). The starting and termination points are always chosen as on and off ramps on the freeway. The following heuristics are adopted to determine the starting and termination points of the diversion:

- If the incident severity index is equal to 3 or 4, employ a freeway-freeway diversion strategy.
- If the incident is near an exit ramp of a directed link, divert at least two exits prior to the incident link.
- Assume that only one entrance/exit combination will be employed for the diversion, then find the route within the search area that is closest to minimizing the sum of arterial and freeway delays.

7.8.2 Estimation of Diversion Volume

An important element of the diversion strategy that needs to be determined is the volume that is to be diverted. The volume is needed to predict, roughly, the v/c ratios on the network links after the diversion is initiated. In the previous implementation of the network generator, the diversion volume was a predefined fraction of the existing freeway link volume [15]. That approach is simplistic and is refined in the enhanced version of the network generator described in [14].

Gupta et al. [8] approach the problem of estimating the diversion volume by regarding any volume that produces a *v/c* ratio greater than a pre-specified critical threshold as the potential volume for diversion. A decision is then made to check if the excess volume when diverted reaches the termination location of the diversion faster than if it had not been diverted. If so, the diversion is implemented. They also investigated whether diversion is beneficial for prespecified such *v/c* thresholds [8].

The approach the network generator follows is analogous to that approach in [8] but has several important advantages. To illustrate those advantages, consider a major incident that makes the *v/c* ratio very high (could be much greater than 1). In the system of [8], the option of reducing the *v/c* ratio to one of the few prespecified thresholds (all less than 1) would entail the diversion of either very large amounts of link volume or none at all. That approach does not allow for consideration of a larger set of possible diversion volumes from which could be selected the most beneficial diversion. One way to consider a larger set of diversion volumes would be to consider a large set of *v/c* ratios (both less than and greater than 1). Instead, the network generator proposes to accomplish that goal through a more intuitive approach in which the diversion volume is first set to an online generated lower bound and is then continually marginally increased until the best diversion volume is identified. The best strategy will then be identified in terms of travel time increases due to the extra volume that is going to be diverted.

The diversion volumes have to be bounded between zero and the link volume on the freeway. However, to reduce the search space for the diversion volume, certain upper and lower bounds for the diversion volume need to be set. The value for those bounds can be fixed based on a variety of factors. The factors that need to be considered for the lower bound on the diversion volume are mainly the incident duration and the available capacity. The upper bound for the diversion volume is also fixed based on the severity of the incident. Note that if the arterials are very congested or if the incident is not very severe, the upper bound on the diversion volume can be set to be much less than the entire link volume. Thus, the network generator proposes an iterative stepwise search between online generated lower bounds and upper bounds for the diversion volume. The lower and upper bounds are expressed as fractions of the freeway volume at the link corresponding to the exit ramp where the diversion route begins.

If the most beneficial diversion strategy is found to be at the lower bound, then we can decrease the lower bound and continue searching in the direction of the new lower bound. Similarly, if the best diversion strategy is found to be at the upper bound, then we increase the bound and continue searching in the direction of the new upper bound. That is a quick, heuristic procedure to roughly estimate v/c ratios which is similar to the incremental assignment technique used to obtain an equilibrium solution for traffic assignment problems. However, the heuristic search process can be time consuming and can considerably delay the working of the system. Thus, it should be used mainly off-line to generate different v/c ratios for different conditions rather than for real-time diversion situations.

7.8.3 Dynamic Link Elimination Concept

An urban arterial network typically comprises a large number of links with different traffic and geometric characteristics. Any route determination model, for diversion or for dynamic traffic assignment, must consider all the links to be potential components of the route that is to be determined. The route generation procedures of the diversion or route guidance models often become computationally expensive because they have to consider all the links within a given search area. Often, the route generation models end up considering a large number of links that are not a part of the routes they generate. Thus, modules that use some approach to identify and eliminate links not likely to be a part of those routes would increase considerably the computational efficiency of the route generation/route guidance models. We call such a module a "link elimination module" and the concept "link elimination."

The real-time factors can be traffic conditions, special events, construction, and maintenance activities. The existence of one of those real-time factors at some point can cause a link not to be included as part of a route during that time period. Thus, the link elimination process has to be dynamic and consider time-varying link conditions.

In addition to the network aggregation models described previously, some additional research may be applicable to the development of the link elimination concept. The notion that factors other than the v/c ratio have to be considered is expressed in [7] for diversion route generation. That model also recognizes that the v/c ratios on the alternative routes may vary with time. Although the approach does not directly employ a link elimination module, it uses an alternative route selection strategy that uses several criteria other than the v/c ratio to determine the diversion route. The model, in effect, eliminates a route if either a particular link in the route violates some criterion or if the characteristic of the route as a whole (e.g., route length in terms of travel time) violates some predetermined travel time criterion. It also considers special information such as the presence of schools and special events in predetermining diversion routes.

The approach proposed in this research is also an enhancement over the network aggregation models because it considers dynamic factors for link elimination. In addition, we recognize that the arterial conditions can vary with time, and we also recognize the need to consider factors other than the v/c condition in link elimination.

On the other hand, compared to the approach proposed by [7], the approach proposed in this chapter has the following characteristics:

- We first eliminate links and then build routes. In comparison, the approach proposed by [7] uses predefined routes and then eliminates the route if a link violates some condition. Thus, this new approach is much more direct and flexible.
- As explained in the following sections, a larger and more comprehensive set of factors to decide whether a link can be a potential element of a route is used in this research.
- Instead of choosing from a limited set of predefined alternative routes, new diversion routes are chosen after eliminating infeasible links.

Thus, the network elimination process proposed in this chapter builds on the research presented in [8, 15, 21, 22] and expands it to include several new features.

7.8.4 Proposed Approach for Link Elimination

The approach adopted for link elimination considers a large set of static and dynamic factors that decide the feasibility of a link for diversion/dynamic route guidance. In comparison to the network aggregation models, the analytic processing of the decision factors is performed heuristically, not mathematically. That makes the processing much more computationally efficient, therefore making it much more suitable for real-time implementation.

The heuristics developed examine each link with respect to a set of decision factors. A decision is then made regarding the feasibility of the link. The heuristics are represented as rules in an expert system, and the set of production rules employed for the network generator constitutes a deterministic set of rules. The antecedents and conclusions of those rules are well defined without being fuzzy. The various decision factors considered form the antecedents to the rules. The decision factors considered, the heuristics involved, and the heuristic processing mechanisms are described in detail in the sections that follow.

An earlier version of the network generator employed classical Boolean logic production rules for link elimination [15]. The major obstacle with the representation of knowledge base using only predicate calculus and Boolean logic is that when the knowledge involves uncertain events (e.g., unknown volumes), the knowledge is seldom of a binary nature. In addition, the early prototype expert system did not recognize that different sets of antecedents may lead to different confidences in the conclusions reached. For example, a conclusion to eliminate the link because of a severe HAZMAT spill has a different confidence level compared to a conclusion to eliminate the link because the link is in an undesirable neighborhood. Simple binary logic systems thus are prone to making false positive and false negative predictions because of a lack of probabilities being associated with the derived conclusions.

To overcome the difficulty of incorporating probabilistic knowledge in binary logic, the enhanced link elimination module uses weights for the degree of belief in the rule. That approach was used with success in the medical diagnostic system MYCIN [23], in which the production rules are weighted by a certainty factor. Positive certainty factors are a measure of relative support for a hypothesis, and negative factors are a measure against the hypothesis.

7.8.5 Factors Influencing Link Elimination

To better understand a decision process that will judge whether a link is feasible for diversion, we may consider the decision to be an outcome of the interaction of external forces and link attributes. We can view any factor that may cause a link to be eliminated as an external force. Link attributes are characteristics of the link that deal with external forces. The decision process then depends on our prediction of the response of the link, which depends on its characteristics, to the external forces. If the expert believes the link is capable of handling the external forces well, then we do not eliminate the link. Tables 7.4 and 7.5 list the external factors and link attributes that the network generator considers in deciding whether a link will be eliminated.

A wide array of decision factors can result from a combination of different external forces that can vary in their extent, magnitude, and link attributes. The different factors considered by the network generator are listed as the antecedents of the heuristic rules developed for dynamic link elimination. At that junction, however, discussions of all the external forces and the magnitudes they can assume and the link attributes and the values they can take are appropriate. The justification for the choices of external forces and link attributes is also illustrated with an example.

7.8.5.1 External Forces

- *Incident severity.* Incident severity affects many of the decisions about using a link. If severity is very high, we may compromise on many other factors in deciding the feasibility of a link. As an illustration, if

Table 7.4
External Forces on a Link [14]

External Forces
Incident Severity: IncidentType and Incident Delay
Traffic Volumes
Weather Conditions
Jurisdictional Issues
Special Events
Construction/Maintenance
Undesired Neighborhoods
Office/School Times
Incidents on neighboring arterials
Adjacency to Hospitals / Metros

Table 7.5
Link Attributes [14]

Link Attributes
Capacity
Geometric Design
Ice/Snow Clearance
Speed Limit
Link Location
Link Type

the incident delay is very high, we may not want to avoid a link despite its speed limit being low.

- *Traffic volumes.* Traffic volume represents the net volume on the link. That includes the volume that normally flows on the links and the additional volume added by the diversion. Traffic volumes can be thought of as external forces because they are an input to determining the congestion level of a link. If the congestion level of the link is high, we may need to eliminate the link.

- *Weather conditions.* Weather forces are clearly external forces. These forces can prevent use of the link for diversion, even when the link is not congested. While quantifying this external force completely can be a difficult task, the heuristics for the network generator work with descriptive measures for the weather. For example, very icy is a descriptive measure of the weather. Table 7.6 lists the different classifications of weather conditions that the network generator recognizes.

- *Jurisdictional factors.* Sometimes jurisdictional factors may not permit some of the links in the arterial network to be used for diversion. The links that have to be eliminated for diversion depend on the location of the incident.

- *Special events.* Special events, like football games, can generate additional traffic on the arterial network. The volumes generated may be so high that they will not permit additional diverted traffic.

- *Construction and maintenance.* In the event of maintenance activity on some arterial links, the link may be deemed unfit for diversion. In the modeling of the network generator, the operator has to input the links that are currently undergoing maintenance.

- *Unsafe neighborhoods.* One of the important factors that affect the compliance of users to the diversion recommendations is the type of neighborhoods through which the diversion routes pass. It has been shown through some preliminary observations that drivers are less willing to divert if the diversion route goes through an unsafe neighborhood.

- *School and office hours.* This external force is similar to special events except that it has to be dealt with every day.

- *Arterial incidents.* Incidents on arterial streets are an external force that may cause a link to be eliminated for diversion of traffic from freeways.

Table 7.6
Classification of Weather Conditions [14]

Snow	Ice	Rain	Fog
"Light Snow"	"Icy"	"Light Rain"	"Heavy Fog"
"Moderate Snow"	"Very Icy"	"Heavy Rain"	
"Heavy Snow"			

- *Hospitals, airports, metros.* Often in urban cities, it may not be advisable to divert traffic onto streets adjacent to hospitals, airports, or metros. That is because the traffic disruptions brought about by the diverted volumes may prevent easy access to those locations.

7.8.5.2 Link Characteristics

- *Link capacity.* Link capacity is an important characteristic that the diversion model needs to consider. The effect of loading a given diversion volume can be determined only with a knowledge of the link capacity.
- *Geometric characteristics.* Geometric characteristics of major roads are particularly important in bad weather. Consider the example of a road with a steep gradient. It may not be appropriate to divert traffic onto such a road when the weather is icy. In addition, some roads can be prone to incidents under heavy congestion due to their geometric characteristics. The values this property may take are listed in Table 7.7.
- *Ice and snow clearance.* This property of a link is needed to determine its feasibility when the weather is icy or snowy. If the link is periodically cleared, we may not want to eliminate it. If the link is not cleared, then we may want to eliminate it. The values this property can take are listed in Table 7.8.
- *Speed limit.* A low speed limit may cause a link to be eliminated because drivers may not be willing to divert onto such a link.
- *Location.* The location of a link with respect to undesireable neighborhoods, office and school zones, special events, and airports influences its elimination.
- *Link type.* Link type has an influence on our decisions. The conditions under which we eliminate an arterial link may be different from the conditions under which we eliminate a freeway link or an entrance or exit ramp. Types of links are listed in Table 7.9.

Table 7.7
Geometric Characteristics [14]

Geometric Characteristics of Links	
Gradient	Curvature
Steep	Windy
Level	Straight

Table 7.8
Ice and Snow Clearance Measures [14]

Link Clearance Measures
Adequately Cleared
Moderately Cleared
Not Cleared

Table 7.9
Types of Links [14]

Link Types
Arterial Street
Freeway Link
Entrance Ramp
Exit Ramp
Bridge / Tunnel

7.8.6 Rule Base for Dynamic Link Elimination

A rule for dynamic link elimination connects a decision factor with a decision or with a result that could lead to a decision. Before the decision factors are described, the decisions each rule may lead to are described. The possible results a rule may lead to are:

- *Definitely eliminate (DEL).* For some cases, the need for eliminating a link can be strong. In such a case, the link may be considered to be "definitely eliminated." Such a case may occur when safety reasons force the model to definitely eliminate the link.

- *Eliminate (EL).* Sometimes there may not be enough reason to definitely eliminate the link, in which case "eliminate the link" means that the link preferably is not to be used. Links that are eliminated and not definitely eliminated may have to be used in the event that no diversion route is found.

- *Don't eliminate (NE)/definitely don't eliminate (DNE).* If the link does not satisfy the criteria that prompt its elimination, the decision state will be "don't eliminate." "Definitely don't eliminate" rules represent the strong need for keeping a specific link. That can be a freeway link that does not have an alternative for creating another diversion route.

A parallelism can be drawn here between the production rules of the network generator and those of MYCIN [23]. MYCIN's rules serve to diagnose a patient's illness. The diagnosis could be any of a set of possible ailments of varying severity. The network generator's link elimination rules can be similarly thought of as diagnostic rules that ascertain whether a particular link is "healthy" for use in diversion. The network generator's diagnosis for a particular link could be a malfunction that cripples the link's functioning in which the link is definitely eliminated due to icy conditions on a bridge. Or the diagnosis might call for temporary suspension from activity until absolutely necessary, in which case the link is eliminated due to jurisdictional problems. If the link's functioning is normal, the link will not be eliminated, or the link could be diagnosed to have some salient features (a very low v/c) that prompts its definite use.

At this point, a difficulty in developing a rule-processing scheme that will utilize the above decision states may arise because two different rules may be applicable to the same condition and may yield different decision states. We then need to resolve the conflict between the decisions that a set of rules may yield. Extending the MYCIN analogy further leads to a similar case in which a patient has symptoms characteristic of several ailments. In comparison, the prototype expert system described in [15] has link elimination rules with just one decision state, "eliminate." If any one rule's antecedent matched the given condition, the link would be eliminated. While that approach does not have any conflict resolution problems, it leads to some other serious drawbacks. Tests revealed that too many links were often being eliminated because the combination of external forces and link attributes often matched at least one rule antecedent [15, 16]. In the enhanced model, we aim for more realistic outputs and thus adopt a multivalued decision state for our rules and also employ a conflict resolution technique.

7.8.7 Link Elimination Decision Making

The link elimination decision-making task can be represented by the Venn diagram in Figure 7.5. The diagram shows four mutually exclusive sets representing the four decision states discussed in the Section 7.8.6. Each set has characteristics that uniquely identify the decision state with that set. For

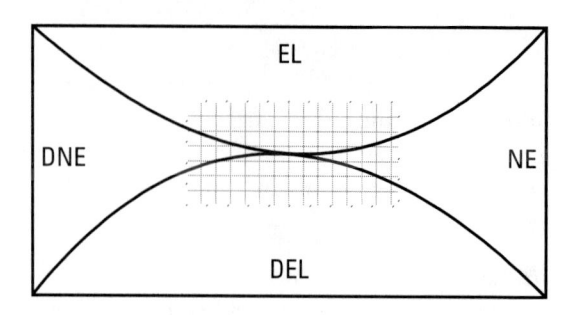

Figure 7.5 Venn diagram for decision states [14].

Legend:
DNE: Definitely do not eliminate
NE: Don't eliminate
EL: Eliminate
DEL: Definitely eliminate
▨ Incidence

example a v/c ratio of less than 0.3 will belong only to the set "Definitely do not eliminate (DNE)," since a link with a v/c ratio less than 0.3 is a very attractive candidate for carrying extra traffic. The combined incidence (external forces + link characteristics) can also be represented by the combined set I. Note that combined set I can share common areas with the other four sets. For example, the combined incidence can consist of a low ratio and a HAZMAT spill. The low v/c ratio belongs to DNE, and HAZMAT spill belongs to the DE set. In this case the incidence shares common areas with the DE and DNE sets.

The decision-making task for the link elimination module is to map a given incidence set to the most appropriate one of the four decision states. To fully comprehend the decision-making process, it is important to discuss the link elimination rule structures in more detail.

7.8.8 Link Elimination Rule Structure

The production rules for the link elimination module incorporate probabilistic knowledge in binary logic by weighing each rule with a confidence factor. Those rules can be represented by:

$$E \rightarrow H$$

Where H is the hypothesis and E is the evidence relating to the hypothesis. For example, a hypothesis might be that safety reasons require a link to be eliminated, and the evidence could be a severe HAZMAT spill around the area.

The network generator rules recognize that the relationships between evidence and hypotheses are often uncertain to the extent we can depend on the hypotheses. For example, a hypothesis to eliminate the link based on safety reasons is better supported by evidence of a HAZMAT spill compared to the link being in an undesirable neighborhood. To accommodate the degree of belief in a hypothesis, the network generator uses certainty factors to weigh each rule. The certainty factor represents change in belief about a hypothesis given some evidence. That can be represented by:

$$CF(H|E). \ E \rightarrow H$$

Certainty factors range between −1 and 1. If the certainty factor for a particular rule is positive, that means the evidence increases the belief in the hypothesis of that rule. On the other hand, if it is negative, it implies that the evidence decreases the belief in the hypothesis of that rule. For example, a high v/c ratio increases our belief in the hypothesis that the link needs to be eliminated, while a low v/c ratio decreases our belief. The two rules can be represented as in Figure 7.6, which indicates that the hypothesis to eliminate a link is strengthened by 0.7 if the evidence that $v/c > 0.8$ is true. On the other hand, the belief in the hypothesis is reduced by 0.9 if there is evidence that $v/c < 0.3$.

7.8.9 Link Elimination Decision Process

The decision regarding whether to eliminate a link is dictated by the decision state reached through the processing of knowledge represented as rules. The rules that are adopted follow the structure discussed in Section 7.8.8. This discussion focuses on how the structure of the rules is exploited to arrive at one of the four decision states shown in Figure 7.5.

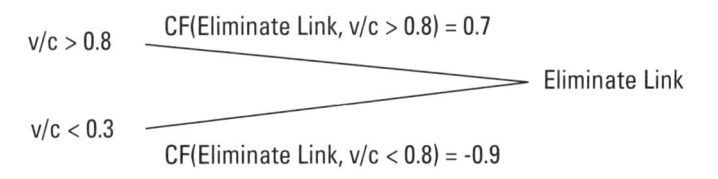

Figure 7.6 Representation of two v/c rules [14].

First, it is important to note that there are some conditions in which a single rule, if applicable, governs the decision state. These rules focus primarily on safety issues (very icy conditions) and obvious feasibility issues (very high congestion during peak periods) during diversion. If any of the rules is applicable, then we definitely eliminate the link. If not, we continue with our processing to inspect the link, with reference to other decision factors.

The critical region that is a subset of the DEL region is shown in Figure 7.7. If the incidence presented has any components that have applicable antecedents in the critical region, the link must be eliminated. The rules that apply for the critical elimination are depicted in Figure 7.8.

If any of the rules presented in Figure 7.8 is applicable, the hypothesis of CRITICAL conditions holds true and the link is definitely eliminated. Once a particular hypothesis is true, we stop processing.

If the incidence does not have a mapping onto the critical region, the rest of the decision-making process described below must be carried out. The philosophy of the decision process rests on the observation that the four decision states may be represented as different measures of belief against an assumed hypothesis. To illustrate that further, let us choose an assumed hypothesis that the link is to be eliminated. It is then possible to write all the rules in such a way that their confidence values are measures of belief against the hypothesis that the link be eliminated. The philosophy of the approach is to use the cumulative

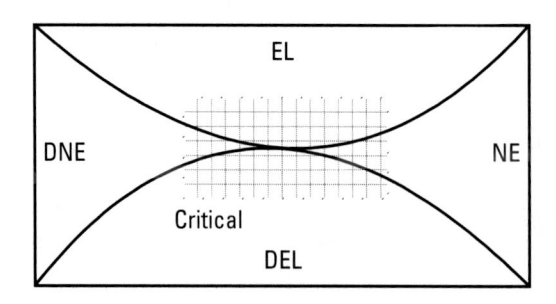

Legend:
DNE: Definitely do not eliminate
NE: Don't eliminate
EL: Eliminate
DEL: Definitely eliminate
▨ Incidence

Figure 7.7 Concept of critical region [14].

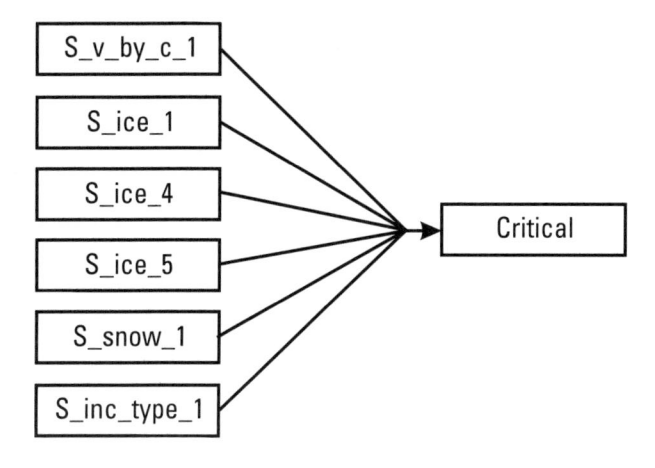

Traffic Diversion for Real-Time Traffic Management During Incidents 199

Figure 7.8 Critical elimination rule network [14].

belief for the incidence against the hypothesis that the link is to be eliminated to map it to the four decision states. The process is represented by Figure 7.9.

If the net confidence against the hypothesis that the link is to be eliminated varies from Cx1 to 1, there is good reason to definitely eliminate the link. If it varies from Cx2 to Cx1, then we may merely eliminate the link. If the net confidence turns out to be negative, we either do not eliminate the link or we "definitely do not eliminate" the link. Thus, if the net confidence varies from Cx2 to Cx3, we do not eliminate the link, and if it is less than Cx3, we definitely do not eliminate the link. Extensive tests and consultation with experts have to be done before the thresholds Cx1, Cx2, and Cx3 are determined.

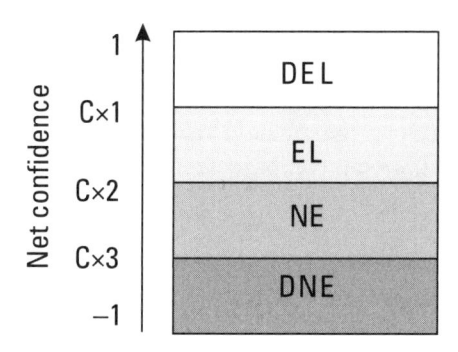

Figure 7.9 Net confidence [14].

7.8.10 Cumulative Weight Function for Conflict Resolution

Again, the cumulative weight function for conflict resolution used for link elimination is similar to one used in MYCIN [23]. The cumulative measure of belief against a hypothesis is computed by the function given by a parallel combination of evidence also given in Kruse et al. (1991):

$$CF(H|E_1, E_2) = \begin{cases} \frac{CF(H|E_1) + CF(H|E_2)}{1 - \min\{|CF(H|E_1)|, |CF(H|E_2)|\}} & \text{if } CF(H|E_1) \cdot CF(H|E_2) \in (-1, 0] \\ \{(CF(H|E_1) + CF(H|E_2)) + (CF(H|E_1) \times CF(H|E_2))\} & \text{if } CF(H|E_1) \text{ and } CF(H|E_2) > 0 \\ \{(CF(H|E_1) + CF(H|E_2)) - (CF(H|E_1) \times CF(H|E_2))\} & \text{if } CF(H|E_1) \text{ and } CF(H|E_2) > 0 \\ \text{undefined} & \text{if } CF(H|E_1) \times CF(H|E_2) = -1 \end{cases} \quad (7.1)$$

It is important to note that the combination function meets the desired condition of not having the cumulative weight being influenced by the order in which the evidence is presented, for parallel combinations.

For sequential combinations of evidence for two rules represented as with sequential combination is performed using the function given in [23] and Kruse et al. (1991)

$$CF(H|E') = CF(H|E) \cdot \max(0, (CF(E|E'))) \quad (7.2)$$

The rule network for decision making is depicted in (1). The decision state finally chosen depends on the cumulative weight. To compute the cumulative weight, all possible rule antecedents have to be evaluated. Thus, each rule that is applicable in the left column is evaluated, and the corresponding influence on the cumulative weight is then computed. Once all the rules are evaluated, the final decision is based on the net measure of belief against the hypothesis of not eliminating the link.

The rule-processing procedure seemingly requires that all rules be evaluated. However, such a case can be avoided by evaluating the mutually exclusive antecedents separately. To illustrate that idea, consider a case in which the weather is icy. We immediately know that snow rules need not be applied. Another illustrative case is the *v/c* ratio test. Instead of checking the *v/c* ratio against each threshold, we can compute it and immediately set the rules that are irrelevant to false.

7.8.11 Rule Antecedents

As explained earlier, a rule antecedent generally comprises a set of external forces coupled with a set of link attributes. If we were to explicitly enumerate all possible antecedents, then a given state of external forces and link attributes could lead to only one possible applicable antecedent and the result of the decision could be obtained immediately. However, the possible antecedents are numerous and do not permit explicit enumeration. That would imply the need for some kind of strategy in which only a limited number of rule antecedents are enumerated and the expert system maps a given combination of external factors and link attributes to several of the enumerated antecedents. If the mapping is unique, then we have a decision. If the mapping is not unique and we have several applicable enumerated antecedents, then we have to use some kind of conflict resolution mechanism.

7.8.12 Link Elimination Rules

The set of link elimination rules can be further classified into simple rules and compound rules. Simple link elimination rules have antecedents that comprise one external force and one link attribute. This section gives a sample of the link elimination rules that the network generator considers (Table 7.10). Each rule is followed by an explanation.

7.9 Route Generation

The network generator has two sets of options for route generation. The first option available is predefined diversion plans. The plans could be a set of diversion routes that are currently being employed by the northern Virginia incident management personnel [2]. If a given incident link has a predefined route associated with it, the network generator has the capability to extract that information and display it. If, on the other hand, such a plan is not available or is not implementable due to reasons like traffic congestion on the arterials or construction activity on one of the links in the plan, then the network generator has the capability to dynamically generate diversion routes. The links that are not eliminated are used to form the routes. The route generation is done through use of a static shortest-path algorithm that generates the shortest route from a given starting point of the diversion to a given termination point. The route generation function of the network generator can be enhanced by using some of the following shortest-path algorithms:

- Static and time-dependent kth shortest-path algorithms. These algorithms have been programmed in C as part of the WAIMSS and can be used as part of network generator.

Table 7.10
Examples of Link Elimination Rules [14]

Hypothesis Name	Antecedent and Explanation	Decision/Action
S_v_by_c_1	*If (volume/capacity > threshold_1) Then* Link is too congested.	DEL
S_v_by_c_2	*If (threshold_1 > volume/capacity > threshold_2) Then* Link is congested so eliminate it.	EL
S_v_by_c_3	*If (threshold_2 > volume/capacity > threshold_3) Then* Link is not so congested and we'd like to retain it for diversion purposes.	NE
S_v_by_c_4	*If (threshold_3 > volume_by_capacity) Then* Link offers excellent potential for diversion, as it is not at all congested.	DNE
S_SPE_1	*If (link location = adjacent _to_ special _event)&(time= special _event _beginning _or _ending _time) Then* This rule takes into account the traffic disturbances created by the continuous stream of traffic entering the arterials.	EL
S_SPE_2	*If (time = special event begin time) Then* If we don't have real time data on link volumes, then we need to account for the extra volumes generated by the special event.	EL
S_ice_1	*If (weather = icy) and (link type = bridge) Then* Bridges can be very dangerous under icy conditions and heavy traffic.	DEL (CRITICAL)
S_ice_2	*If (weather = icy) and (Ice_Snow clearance = adequate) Then* Not many roads in the arterial network will be adequately cleared by the authorities. Hence it may be appropriate not to eliminate any roads that are well maintained.	DNE

Table 7.10 (continued)

Hypothesis Name	Antecedent and Explanation	Decision/Action
S_ice_3	If (weather = icy) and (Ice_Snow_Clearance = moderate) Then	NE
	If diversion is really needed, it might be necessary not to eliminate links that are moderately maintained.	
S_ice_4	If(weather = icy) and (Ice_Snow_Clearance = not cleared) Then	DEL (CRITICAL)
	In such conditions, the link will be a safety hazard.	
S_ice_5	If (weather = very_icy) Then	DEL (CRITICAL)
	In such conditions, diversion must not be initiated. This is used as a double check.	
S_ice_6	if (weather = icy) and (gradient = steep) and (Ice_Snow_Clearance = moderate) Then	EL
	Steep slopes are dangerous in icy conditions.	
S_snow_1- S_snow_5	Similar to S_ice_1 to S_ice_5	EL
S_rain_1	If (rain = heavy_rain) and (road is winding) Then	EL
	This situation makes it very difficult for the diverted traffic to flow smoothly).	
S_fog_1	If (fog = heavy) and (road is winding)Then	EL
	This situation makes it very difficult for the diverted traffic to flow smoothly.	
S_CM	If (link location = Construction/Maintenance Location) and (time = construction period) Then	EL
	The maintenance activity will reduce capacity and also cause additional traffic disturbances.	
S_Bad_Neig	If (link location = Undesirable Neighborhood Zone) and (Time = Night) Then	DEL
	We do not wish to divert through a bad neighborhood.	

Table 7.10 (continued)

Hypothesis Name	Antecedent and Explanation	Decision/Action
S_Sch_Off	If (link_location = School_Office_Location) and (time = office_school_open_close_time) Then We may not wish to add further traffic to this link.	EL
S_Sch_Hospital_Metro_Air	If (link_location = Hospital/Metro/Airport location) Then We don't wish to add more traffic and make access to these sensitive points.	EL
S_Inc_Type_1	If (incident_type = hazmat_severity_1) and (link_location within hazmat_impact_area). Then We don't wish to add more traffic and make access to these sensitive points.	DEL (CRITICAL)
S_Inc_Type_2	If (incident_type = hazmat_severity_2) and (link_location within hazmat_impact_area). Then HAZMAT accidents.	EL
S_adj_bl	If (adjacent links are all DEL) Then If all adjacent links are definitely eliminated, then we eliminate this one too.	DEL (CRITICAL)
S_Juris	If (jurisdictional_factor = yes) Then The reason we don't definitely eliminate the link is that in some cases, it may be possible to override jurisdictional issues by the operator by contacting the appropriate authorities.	

- Static and time-dependent disjoint pairs of shortest-path algorithms that have been developed at the Virginia Tech Center for Transportation Research as part of the WAIMSS research efforts [24].

The use of the enhanced shortest-path algorithms significantly improves the route choice mechanism of the network generator. However, the algorithms also increase computational overhead associated with them, which is an important shortcoming, especially for real-time operations. Thus, the use of advanced static and dynamic shortest-path algorithms has to be tested for large networks before they are put to use for real-time applications.

7.10 Summary and Need for Further Research

This chapter presented some of the more practical studies and projects that deal with the diversion problem during incidents. The network generator, an expert system approach adopted by WAIMSS for determining possible diversion routes in real-time, is discussed in detail. Several research areas need to be explored, as described next.

7.10.1 Route Prioritization

The concept for prioritizing multiple diversion routes determined by the network generator has not yet been implemented and tested. In places such as northern Virginia, multiple diversion routes generally are not available in general. In addition, the dissemination of multiple route information is problematic. Therefore, we briefly present the concept of route prioritization.

Several researchers have estimated probabilistic functions for driver route choice decisions [11, 13]. One way of prioritizing routes is to use those functions once the routes have been generated based on an impedance function criterion, such as travel time. However, the route choice models consider several variables regarding the class of motorists, including household and economic information. While such models are likely to produce accurate predictions on route choice once the data are available, employing those models for real-time diversion is not generally recommended since the required data usually are not known. Hence, we propose an approach in which the route prioritization module of the network generator works by incorporating several travel disutility components in the link impedances used by the shortest-path algorithms of the route generation module.

The factors considered in the travel impedance are of particular interest to diversion. They are chosen based on a study conducted by the Texas Transportation Institute to examine driver propensity for diversion [11].

- Average distance from the freeway of all links in the route;
- Total arterial travel length;
- Travel time on the link.

In addition to those factors, we can also add:

- Additional term for links not on familiar diversion routes;
- Turn penalties at each intersection.

All those factors are combined to give the following form:

$Z = k_1$ (distance from freeway) $+ k_2$ (arterial travel length) $+ k_3$ (travel time on link) $+ k_4$ (penalty for link not being on familiar route)

We can normalize the weights to be of the form

$k_1 + k_2 + k_3 + k_4 = 1$

For intersections, the turning movements can be represented as links where the only nonzero-dependent variable in the impedance function is the travel time.

7.10.2 Testing and Validation of Diversion Strategies

Diversion rule bases similar to the one developed as part of network generator and other diversion algorithms have to be tested using online data before being put to actual use. Some of the diversion models described in the literature have been tested using simulation programs [25, 26]. However, simulation is just another abstraction of reality, and there is no guarantee that it reflects the real route choice behavior of drivers. Moreover, some of the practical effects of weather, link geometry, and so on are not adequately modeled in any of the existing traffic simulation packages. Thus, it is not possible to accurately assess the effects of those fuzzy decision-making variables that are in fact the major decision-making criteria for the incident management personnel, using the existing traffic simulation packages. As a result, there is need for online testing and validation of the diversion models that are being developed mostly in laboratory settings.

7.10.3 Multiple-Point Diversion

Most of the diversion plans and algorithms are designed to deal with point diversion. That is due to the lack of well-tested and reliable algorithms as well as the problems related to real-time information dissemination. A considerable amount of research is needed to develop and implement multiple point diversions in real time.

7.10.4 Network Connectivity and Existence of Multiple Diversion Routes

One of the major problems with the link elimination approach is the loss of the connectivity between two points. By eliminating infeasible links, we might not be able to find any routes between the desired origin-destination pairs. The effects of loss of network connectivity and the ways to prevent it while eliminating infeasible links should be studied. In many cases, it is also desirable to determine more than one diversion route. So far, *k*th shortest-path algorithms and rather simplistic disjoint path determination techniques have been used to determine multiple diversion routes. More research similar to the one presented in [24] is needed to develop reliable and theoretically correct shortest-path algorithms to determine multiple shortest paths for diversion.

Review Questions

1. Obtain a map of the transportation network in your area. Assume that a severe accident happened on one of the major freeways. Determine the possible diversion routes to bypass the incident on the freeway. Discuss the reasons for your diversion route selection(s).

2. For the diversion route(s) selected in question 1, determine the necessary measures for actually implementing a diversion plan. Include the determination of the number and location of VMSs, installation of detour signs, and retiming signals on the arterials.

3. Based on your driving experience in the area, estimate the percentage of drivers who will divert onto the selected routes. Using simple deterministic queuing analysis, determine the queue lengths with and without diversion.

4. Review the link elimination rules of the network generator. Discuss which rules are applicable to your area. Suggest additional rules that might be useful or changes to the threshold values.

References

[1] Virginia Tech Center for Transportation Research, Northern Virginia Incident Database, 1994.

[2] Virginia Department of Transportation, "Northern Virginia Freeway Management Team Operating Manual—A Regional Plan for Traffic Management on Northern Virginia Freeways," 1990.

[3] Real-Time Adaptive Control Strategies (RT-TRACS), available at http://www.rt-tracs.org/.

[4] Hall, W. H., *Research on Route Diversion Parameters*, Final Report prepared for FHWA, Dec. 1974.

[5] Berger, C. R., R. L. Gordon, and P. E. Young, "Single Point Diversion of Freeway Traffic," *Transportation Research Record*, Vol. 601, 1976, pp. 10–17.

[6] Roper, D. H., R. F. Zimowski, and A. M. IWAMAS, "Diversion of Freeway Traffic in Los Angeles: It Worked," *Transportation Research Record*, 957, pp. 1–5, 1984.

[7] Ketselidou, Z., *Potential Use of Knowledge-Based Expert Systems for Post-Incident Traffic Control*, Ph.D. Thesis, Univ. of Massachusetts at Amherst, Feb. 1989.

[8] Gupta, A., V. Maslanka, and G. Spring, "Development of a Prototype KBES in the Management of Congestion," Preprint #920302, Transportation Research Board, Washington, D.C., 1992.

[9] Taylor, M. A. P., "Knowledge-Based Systems for Transport Network Analysis: A Fifth Generation Perspective on Transport Network Problems," *Transportation Research*, Vol. 24A, 1990, No. 1, pp. 3–14.

[10] Cascetta, E., A. Nuzzolo, L. and Biggiero, "Analysis and Modeling of Commuters' Departure Time and Route Choices in Urban Networks," Second Internat. Capri Seminar on Urban Traffic Networks, July 1992.

[11] Ullman, G. L., C. L. Dudek, and K. N. Balke, "Effect of Freeway Corridor Attributes Upon Motorist Diversion Responses to Travel Time Information," 73rd TRB Ann. Meeting, Washington, D.C., Jan. 1994.

[12] Mahmassani, H., C. Caplice, and C. Walton, "Characteristics of Urban Commuter Behavior: Switching Propensity and Use of Information," Preprint #890738, Transportation Research Board, Washington, D.C., 1990.

[13] Khattak A. J., F. S. Koppelman, and J. L. Schofer, Stated Preferences for Investigating Commuters' Diversion Propensity," 71st TRB Ann. Meeting, Washington, D.C., 1992.

[14] Kachroo, P., K. Ozbay, Y. Zhang and W. Wei, "Development of a Wide-Area Incident Management Expert System," (work order #DTFH71-DP86-VA-20) FHWA Final Report, 1997.

[15] Ozbay, K., A.G. Hobeiken, S. Subremanian and V. Khrishnaswamy, "A Heuristic Network Generator for Traffic Diversion During Non-Recurrent Congestion," Transportation Research Board, 73rd Ann. Meeting, Washington, D.C., Jan. 1994.

[16] Krishnaswamy, V., *Heuristic Network Generator—An Expert System Approach for Selection of Alternative Routes During Incident Conditions*, Master's thesis, Virginia Polytechnic Institute and State University, Blacksburg, VA, 1994.

[17] Subramaniam, S., Y. Zhang, *Wide-Area Incident Management Expert-GIS System Development*, Project Progress Report submitted to FHWA, July 1994.

[18] Mastbrook, S., "A Framework for Predicting Urban Freeway Delays in Real-Time," M.S. thesis, Virginia Polytechnic Institute and State University, Blacksburg, VA, 1996.

[19] Morales, J. M., "Analytical Procedures for Estimating Freeway Traffic Congestion," *Public Roads*, Sept. 1986, Vol. 50, No. 2, pp. 55–61.

[20] Bovy, P. H. L., and G. R. M. Jansen, "Spatial Aggregation in Traffic Assignment: Effects of Zone Size and Network Detail Upon All-or-Nothing Assignment Results," 1978 PTRC Summer Ann. Meeting, Warwick, England.

[21] Eash, R. W., et al., *Equilibrium Traffic Assignment on an Aggregated Highway Network for Sketch Planning*, Univ. of Illinois at Urbana Champaign, Transportation Planning Group, Dept. Civil Engineering, 1983.

[22] Haghani, A. E., and M. S. Daskin, "Network Design Application of an Extraction Algorithm for Network Aggregation," *Transportation Research Record*, 944, 1983.

[23] Shortliffe, E.H, and B.G. Bucheran, "A Model for Inexact Recording in Medicine." *Mathematical Biosciences 23*, 1975, pp. 351–379.

[24] Sherali, H. D., K. Ozbay, and S. Subramaniam, "Time-Dependent Shortest Pair of Disjoint Paths: Complexity, Models, and Algorithms", *Networks J.*, Vol. 31, 1998, pp. 235-272.

[25] Cragg, C. A., and M. J. Demetsk, *Simulation Analysis of Route Diversion Strategies for Freeway Incident Management*, draft final report, Virginia Research Council, Charlottesville, 1994.

[26] Reiss, R. A., N. H. Gartner, and S. Cohen, "Dynamic Control and Traffic Performance in a Freeway Corridor: A Simulation Study," *Transportation Research-A*, Vol. 25A, No. 5, 1991, pp. 267-276.

Selected Bibliography

Kluse, E. Schweche, and J. Heinsohn, *Uncertainty and Vagueness in Knowledge Based Systems*, Springer Verlag, Berlin, Heildelberg, 1991.

Mahmassani, H. S., et al., *Development and Testing of Dynamic Traffic Assignment and Simulation Procedures for ATIS/ATMS Applications*, Technical Report DTFH61-90-R-00074-FG, 1994.

Ziliaskopoulos, A., and H. Mahmassani, "Time-Dependent Shortest Path Algorithm for Real-Time Intelligent Vehicle Highway System Applications," *Transportation Research Record*. 1408, 94-100, 1993.

8

Online Traffic Control

8.1 Introduction

Chapter 7 discussed the use of diversion during major incidents. One of the important problems for the effective use of diversion is the question of real-time online traffic control. This chapter gives a brief summary of on-line traffic control issues with a special emphasis on the use of feedback control techniques for on-line traffic control.

The interest in ITS stems from the great advances that have been made recently in the area of sensors, microprocessors, and actuators. For instance, pre-timed signal control algorithms used on signalized traffic intersections cannot respond to the changing arrival patterns of vehicles. On the other hand, with the use of traffic sensors, the traffic queues can be monitored in real time, and traffic signals can be made responsive to the changing conditions of the traffic. That would be similar to having a traffic police officer controlling traffic based on the visual input of traffic conditions.

Traffic can be viewed macroscopically or microscopically. In the macroscopic setting, we view traffic in terms of aggregate variables such as traffic density and traffic volume. Microscopic entities in traffic are the individual vehicles that make up the traffic. There are various control design problems in both settings. The focus of this chapter is twofold:

- Description of the on-line traffic control within the context of ITS;
- Description of a feedback control technique for dynamic traffic routing/control during major incidents.

8.2 Traffic Control Problems in ITS: Dynamic Traffic Routing/Assignment

Dynamic traffic routing/assignment (DTR/DTA) has been an important research topic in transportation engineering [1-7]. DTR specifically refers to the process of diverting traffic at a diversion point dynamically. Static diversion would be the case when the amount of traffic to be diverted has been pre-calculated and does not change with time. Being dynamic implies that the values change with time as traffic conditions change. Figure 8.1 shows a sample site where dynamic traffic routing would be highly beneficial. At the traffic site, which is in Virginia, most of the traffic travels from point A to point B in the morning rush hours and in the reverse direction during the evening hours. Two highways are connected at point A. Vehicles that want to go to point B can take either of the two highways, and normally they take comparable amounts of time. If there is congestion in one of the routes, the travel time on that route increases. Hence, more traffic should be diverted onto the other route. In general, if the travel time is same on both the routes, we can claim that the traffic system is working well. That objective is called user equilibrium, because users try to or would like to emulate that kind of route choice behavior to obtain maximum benefit. Another objective would be to obtain "system optimal," which means that for the total traffic network the overall time created by using the specific choice of traffic diversions at all traffic nodes is optimal (which implies it is less than the total travel time created by any other choice of diversion strategy).

Although there have been several recent attempts to develop real-time DTA/DTR algorithms that will perform on-line traffic control, the majority of the transportation research has focused on off-line planning problems that can be categorized as open-loop approaches. With the advent of ITS, the need for dynamic models, capable of working in real time has become clearer. The traditional optimization-based open-loop approaches attempt to solve the DTA/DTR problem by optimizing the objective functions for the nominal model over the "planning horizon." For real-time traffic flow control, where on-line sensor information and actuation methods are available, this technique is not very well suited.

Zhang et al (1995) discusses the differences between open loop and closed loop control in the context of ramp metering. However, some concepts shown in Figure 8.2(a) apply to all kinds of traffic control problems. In open-loop control, the control input, $u(t)$ is independent of the system input, $r(t)$. However, closed-loop control, depicted in Figure 8.2(b), is more sensitive and robust because it responds to the error $e(t)$, calculated using the system input $r(t)$ obtained from its environment. In the area of traffic flow control, many

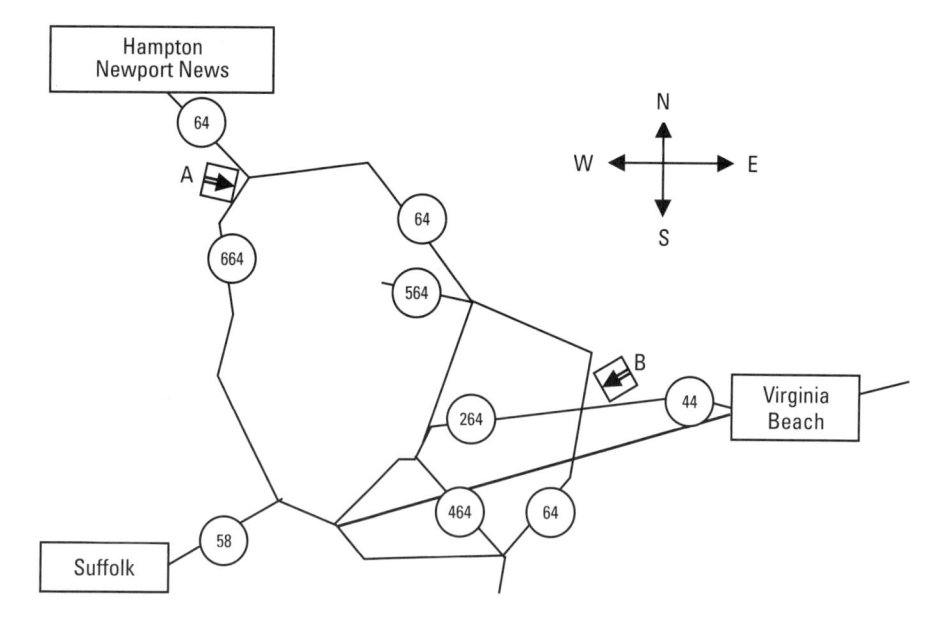

Figure 8.1 A sample site for diversion.

researchers have developed several traffic control models using one of those approaches.

8.2.1 Traditional Techniques

Online traffic management differs from transportation planning in many different ways. A diversion system that will be used for online traffic management should make an effective use of real-time traffic data obtained from sensors. In

(a) Open-loop control

(b) Closed-loop control

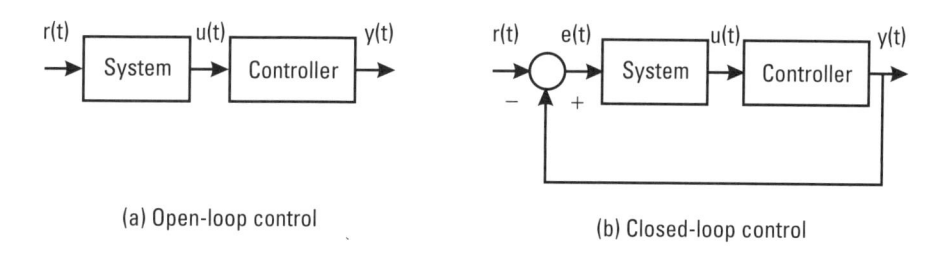

Figure 8.2 Block diagrams of (a) open-loop and (b) closed-loop control.

other words, it should depend heavily on the data to manage the traffic. It should work in real time and be responsive to the changes in traffic flow. Those constraints create a need for a better definition of dynamic traffic routing/control problem. In the light of the facts described above, we can cite the following points for showing that traditional optimization-based approaches to solving DTR/DTA fall short of satisfying the requirement of real-time online traffic management system.

- Optimization-based approaches take the present state of the transportation system that is modeled and optimized and try to optimize the system for the planning horizon. In doing so, they predict future states of the system based on the predefined assumptions about the dynamics of the system. That approach works well for planning purposes in which the time variances of the system do not affect the system too much. For on-line control, however, the classical optimization approach, which depends on the present network conditions and predictions to optimize 1 hour (more or less) into the future, is not realistic.
- The smallest disturbance to the system completely changes the set-optimized predictions determined by the optimization approach. That could mean the extensive computations used to reach the global (or local) optimal solution could be wasted because of a small disturbance.
- The complexity of the optimization models makes it difficult to solve these models in real-time. A solution that does not work in real time is not acceptable at all for online traffic management and control.
- Existing DTR models that optimize the traffic system by distributing traffic flows to the alternative routes do not address at all to the problem of effectively using this information in real time. Once the optimal traffic flows for each time period are determined by one of the existing open-loop DTA/DTR models, the implementation of the results for online traffic control is never discussed in detail. However, this is a major problem because these models are not designed to produce specific control outputs that will generate the desired effects.

In brief, those are the major shortcomings of the traditional techniques for solving traffic control problems using open-loop optimization and optimal control techniques, essentially offline techniques that have been modified for the real-time ITS applications.

There is abundant literature on the traditional techniques for DTR/DTA problems. Among the open-loop modeling techniques adopted for determining optimal traffic flows under certain controls are optimal control, simulation, and mathematical programming techniques. Gartner et al. [8] developed a multilevel control system incorporating local-level and corridor-level controls. The framework recognizes and emphasizes the completion of a loop between the system outputs and the control inputs. However, it relies on the use of static equilibrium algorithms to determine the traffic flows to be diverted. Perhaps the most popular way of solving the DTR/DTA problem is through the use of traditional optimization-based approaches that attempt to optimize the objective functions for the nominal model over the planning horizon [9–11]. Some of the researchers who developed open-loop models using optimal control theory for dynamic traffic assignment problem include [9, 12, 13]. Those researchers employed optimal control theory to develop different models of dynamic traffic assignment problems and used mathematical programming techniques to solve those optimal control formulations. For example, [13] used optimal control theory to develop an equivalent optimization model of the instantaneous user optimal route choice problem. The most important feature of the models and solution algorithms from the online traffic control perspective is that they were not designed to make use of the input received from the real traffic system as a result of the control measures they prescribed.

8.2.1.1 Control DTR/DTA Algorithm Design

Let us now assume the simplest point diversion example, in which a portion of the traffic has to be diverted from the congested route to an uncongested route. The question arises as to how we can achieve the right amount of diversion to equate the travel times on the two alternative routes. One issue is the calculations of the right split factor (i.e., the percentage of traffic flow entering each alternative route compared to the total traffic flow coming to the node). For a real-time traffic responsive system, the split factors should be functions of instantaneous traffic conditions (e.g., traffic densities, flow, or traffic speed at various locations on the routes). For instance, if traffic density generally is more on one route at some time, then instantaneously we could try to change the split factor so that more traffic goes to other alternative routes. Development of algorithms that calculate the split factor values, which are functions of the traffic variables, is the main problem that can be addressed using real-time feedback control algorithms.

A closed-loop algorithm possesses the following elements to be able to function properly.

8.2.1.2 Sensing

To obtain the values of the split factors in real time as functions of traffic variables, the traffic variables need to be measured. Various types of traffic sensors are used for that purpose. Traffic sensors can use various technologies such as video cameras, loop detectors (piezoelectric sensor), and fiber optics. Figure 8.3 shows a traffic camera used in the Transguide system of the Texas DOT (TxDOT).

8.2.1.3 Actuation

After the control algorithm calculates the split factor desired at each traffic node, the split factor value has to be implemented. Various actuation techniques can be employed for real-time implementation of the developed control strategy, including VMSs, HAR, and in-vehicle information.

8.2.1.4 Automatic Control Versus Human-in-the-Loop Control

After being acquired by sensors, the data are processed by computers and can be displayed to operators in a traffic center, who can decide traffic control measures such as VMSs or what signal timing control strategies to use. In automatic mode, some of the functions of the human operators could be bypassed. For instance, if research could clearly indicate how VMSs affect the split factors at traffic nodes, the feedback control algorithm would calculate the split factors automatically based on the measured data and then drive the VMSs. The split factor data can also be used by traffic operators for developing control measures based on those values.

8.2.2 Ramp Metering Control

On of the most effective ways of controlling traffic, especially during incidents, is ramp metering [14]. Ramp metering improves traffic flow by regulating the ramp inflow to a freeway. By effectively controlling the ramp flow, the traffic density on the mainline freeway can be kept below critical level to provide high throughput that is congestion free. For that type of operation, many factors have to be considered such as:

- The inflow at the mainline;
- The queue-holding capacity of the ramp;
- Availability of sensors;
- The arterial system connected to the ramp system.

Ramp metering can help provide a smooth flow of traffic on urban freeways. Moreover, it can also help alleviate congestion on the freeways. The design of ramp metering entails measuring some traffic variables on the freeway and adjusting the ramp metering rate to provide smooth flow. This structure of performing measurements using sensors and in real time adjusting the ramp metering rates renders the problem as that of a closed-loop feedback control problem.

Ramp metering attempts to keep mainline volumes below capacity by controlling the number of vehicles entering the freeway. Under ideal conditions, the wait on the entrance ramp would be compensated for by increased speeds after entering the freeway. Ramp meters can increase freeway speeds while providing increased safety in merging and reducing rear-end collisions on the ramps themselves. The topology of a ramp metering system is shown in Figure 8.3.

Feedback control for ramp metering can be an effective solution for alleviating traffic congestion. The designer of a controller needs to address issues such as controllability and observability of the traffic system, actuation and sensing, robustness, and stability of the closed-loop system. The actuation of the system can be achieved by the light signal, which indicates whether or not vehicles can go onto the freeway. Variables such as traffic density and average traffic speed can be sensed with various types of traffic sensors, such as inductive loops, traffic cameras, and transponders.

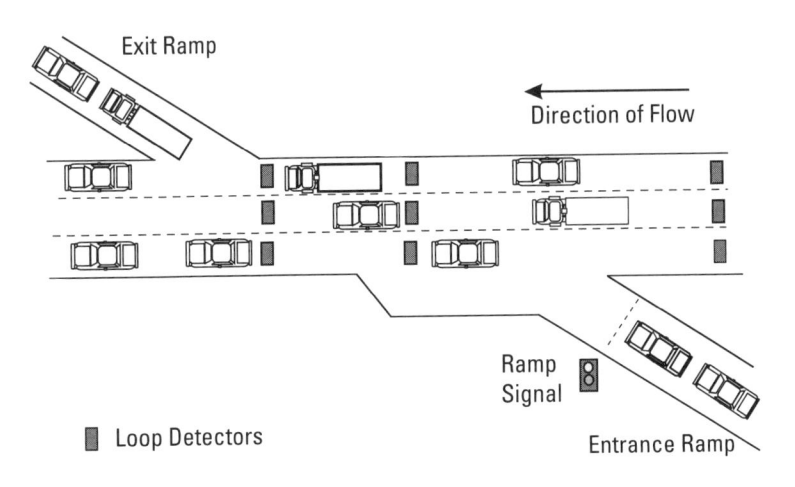

Figure 8.3 Ramp metering topology.

Some of the early work is documented in references [15–22]. That work is related to merge control and ramp metering control design based on demand-capacity relationships. Some early deployment studies were also performed at various sites such as Chicago and Houston. References [20–24] show the work, which used optimization techniques for solving optimal ramp control problems. Some current evaluation studies and simulation-based evaluation methods are described in [23, 24]. Some researchers have designed feedback control laws for ramp metering [11, 25–37]. Those laws are designed after performance of linearization of the dynamics about the nominal equilibrium state. Recently, Kachroo [38] proposed a new scheme for more effective ramp metering. Ramp meters have been deployed in many places internationally, such as the United States [29], France [30], Italy [31], Germany [32], New Zealand [33], the United Kingdom [34], and the Netherlands [35].

8.2.3 Signalized Intersection Control

Diversion of traffic around an incident generally involves the use of neighboring arterials as diversion routes. However, most of the traffic signals on the arterials use predetermined signal plans that function properly under normal traffic conditions. The additional volume due to the diverted traffic requires traffic-responsive signal control. Signalized intersections control the queue dissipation of vehicles in a controlled manner. The actuation is accomplished by traffic lights that provide the signal for controlling the traffic flows in various phases. Currently, many intersections similar to the one shown in Figure 8.4 are being equipped with various sensors to measure the presence of vehicles or the queue lengths in various lanes. Some of the sensors that are being used include inductive loops and camera-based systems. Several feedback control theory-based signal control algorithms have also been proposed in [8, 39].

8.2.4 Traffic Speed Control

Traffic speed on a highway or even on surface streets can be controlled by speed signs. In an ITS, VMSs are used to change posted speeds, based on the prevailing traffic conditions measured by the traffic sensors.

8.3 Feedback Control Designs for Macroscopic Control Problems

This section presents the design of a feedback-based algorithm using a simple point diversion control example. Feedback control, as the term implies, refers

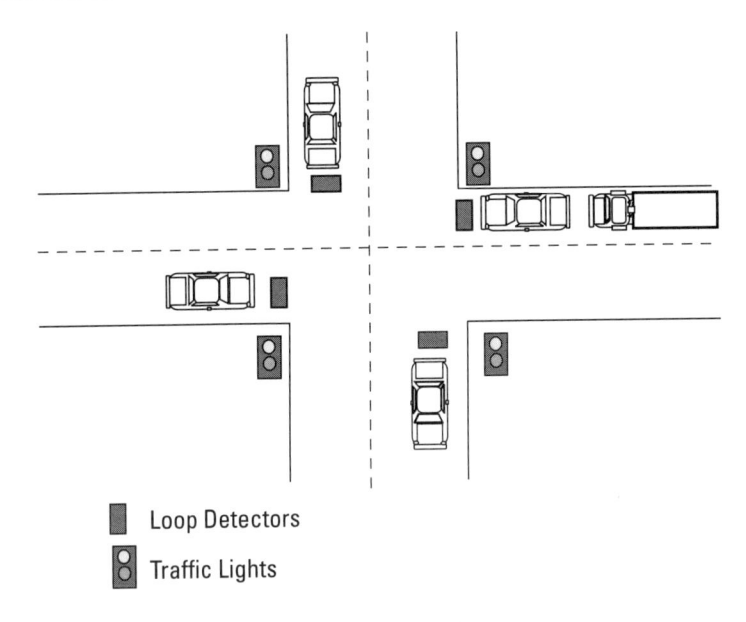

Figure 8.4 Signalized traffic intersection control topology.

to the control of a system to achieve certain specified objectives by feeding some output signal of the system back into the controller. The feedback mechanism makes the system a closed-loop control system. In open-loop systems, there is no feedback. That difference can be illustrated by a simple but representative example of driving a vehicle on a curved track by controlling the throttle and steering. An open-loop control design would start by a mathematically modeling system, in this case, the vehicle, that will be controlled.

An optimal control, which is an open-loop control, can be designed to optimize some objective functional based on this mathematical model assuming the parameters of the model are accurate. For instance, if the model is nonlinear, we could use a nonlinear numerical technique such as a sequential-gradient-restoration algorithm to solve for the control signals. That would give us time-varying control signals for the throttle and the steering from initial time to the final time of the planning horizon. Then, that would mean the vehicle would be driven by the control signals without any real-time input from the environment. This controller is not designed explicitly to deal with the real perturbations to the mathematical model due to the unmodeled dynamics of the actual system. For example, the perturbation could be due to the simplifications during the modeling process or lack of knowledge of some system parameters or dynamics and real-time occurrence of an unexpected situation. Those perturbations could make the system at least suboptimal, if not unstable, since the optimality was calculated based on the nominal offline parameters and initial conditions.

Feedback control design, on the other hand, is based on real-time feedback information. In our example, we could use sensors, which would provide the controller with the real-time information about the coordinates of the vehicle with respect to some convenient frame of reference on the track. The controller would be driven by the difference between the sensed location of the car on the track and the desired location of the car at each sampling instant.

To overcome those problems, we propose a feedback control approach to the DTA/DTR problem. A similar approach has been discussed by Papageorgiou and his colleagues [40] for simple traffic networks consisting of two alternative routes. At every sampling interval, the feedback controller produces control signals that react to the information coming from the sensors. The control update time in this situation is close to the sensor update time, which is of orders of magnitude lower than what would typically be used for a pseudo-online optimal control problem. Because a typical time horizon for an optimal control approach is much higher than the sensor sampling update time, the optimal control solution would be sensitive to the short-time traffic flow changes within that time horizon. The feedback control approach does not have that important limitation. In fact, feedback control can be approximately as responsive as the sensor update sampling time.

Moreover, the resources required to solve an optimal control problem are much more than what are needed for feedback control. That is because the optimal control problem has to be solved for the whole planning horizon, whereas feedback control responds to the sensed variables almost immediately. In the actual implementation of the feedback control, minimal calculations are required to generate the control signal from the sensed variables. For instance, in a proportional-integral-derivative (PID) control of a single-input, single-output (SISO) system, only three additions and one multiplication are required to generate the control signal. Compared to that, the optimal control model requires intensive processing prone to errors due to the lack of sampling within the time horizon.

There are essentially three ways the system modeling could be used to design feedback controllers for macroscopic traffic problems:

- Distributed parameter setting, represented by partial differential equations (PDEs);
- Continuous time lumped parameter setting, represented by continuous-time ordinary differential equations (ODE);

- Discrete time lumped parameter setting, represented by continuous time ODEs.

There are various advantages and disadvantages in designing the feedback controller using any one of the three kinds of models. The original PDE model is derived from the hydrodynamic analogy presented by Lighthill and Whitman [37]. It is, however, difficult to design a feedback controller directly for a distributed parameter system, and that is an area of active research. By space discretizing the model, we can design a feedback controller in the continuous time domain, which can be easier to design. This model obviously will have discretization errors, which could be reduced by designing a robust controller that would attempt to eliminate the errors. Finally, it is natural to design a controller using a discrete time ODE model of the system for discrete implementation of the control. Again, the model would have more discretization errors, and the controller would have to minimize their effect.

Feedback control paradigm in a block diagram can be represented as shown in Figure 8.5. The plant represents the traffic system, whose variables are measured using sensors or observers. There might be a feedforward term in the controller and possibly adaptation based on the input-output data (as shown).

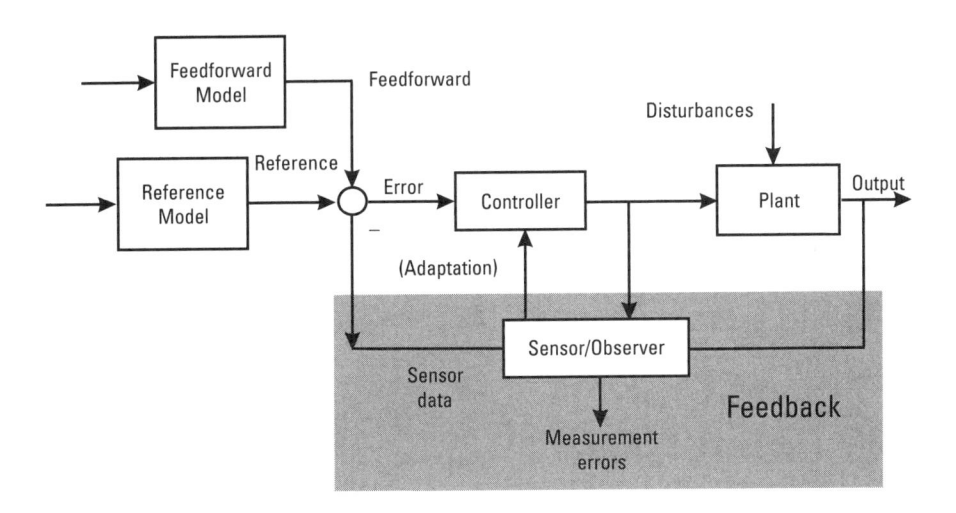

Figure 8.5 Feedback control block diagram.

The idea is to represent the system model in state-pace form and then apply state-space feedback control design techniques. Generally, the problems are nonlinear and contain uncertainties. A strong candidate for solving such problems has been the nonlinear H8 control technique; for some others, feedback linearization; and sliding mode control has proven effective [1–7, 14, 38–40]. The model for a signalized intersection turns out to be a hybrid model using finite state machines with discrete control; however, H8 control also has proved effective in control design for such problem structure.

8.4 Example Problem

To illustrate the ideas discussed here, we have designed a feedback control law for two alternative routes with a single section each (the problem illustrated in Figure 8.1). The control is based on feedback linearization technique for nonlinear systems.

The space-discretized flow equations used for the two alternative routes are:

$$\dot{\rho}_1 = -\frac{1}{\delta_1}\left[v_{f1}\rho_1\left(1 - \frac{\rho_1}{\rho_{m1}}\right) - \beta U\right] \tag{8.1}$$

$$\dot{\rho}_2 = -\frac{1}{\delta_2}\left[v_{f2}\rho_2\left(1 - \frac{\rho_2}{\rho_{m2}}\right) + \beta U - U\right] \tag{8.2}$$

where U is the input flow, β is the split factor, ρ is traffic density, v_f is free-flow section speed, δ is section parameter, and ρ_m is the traffic jam density. Details are given in [3].

We have considered a simple first-order travel time function, which is obtained by dividing the length of a section by the average velocity of the vehicles on it. According to that, the travel time can be calculated as

$$\chi_1(k) = d_1 / \left[v_{f1}\left(1 - \frac{\rho_1}{\rho_{m1}}\right)\right] \tag{8.3}$$

$$\chi_2(k) = d_2 / \left[v_{f2}\left(1 - \frac{\rho_2}{\rho_{m2}}\right)\right] \tag{8.4}$$

where d_1 and d_2 are section lengths, v_{f1} and v_{f2} are the free-flow speeds of each section, and ρ_{m1} and ρ_{m2} are the maximum (jam) densities of each section. We need to equate the travel times on the two sections for optimal desired performance. Hence, we take the new transformed state variable y as the difference in travel times. Differentiating the equation representing y in terms of the state variables introduces the input split factor into the dynamic equation. Therefore, that transformed equation can be used to design the input that cancels the nonlinearities of the system and introduces a design input v, which can be used to place the poles of the error equation for asymptotic stability. These steps are shown next.

The variable y is equal to the difference in the travel time on the two sections.

$$y = \frac{k_1}{(k_2 - \rho_1)} - \frac{k_3}{(k_4 - \rho_2)} \tag{8.5}$$

where

$$k_1 = \frac{d_1 \cdot \rho_{m1}}{v_{f1}}, \quad k_2 = \rho_{m1}, \quad k_3 = \frac{d_2 \cdot \rho_{m2}}{v_{f2}}, \quad k_4 = \rho_{m2} \tag{8.5a}$$

Equation (8.5) can be differentiated with respect to time to give the travel time difference dynamics.

$$\dot{y} = \frac{k_1 \dot{\rho}_1}{(k_2 - \rho_1)^2} - \frac{k_3 \dot{\rho}_2}{(k_4 - \rho_2)^2} \tag{8.6}$$

By substituting (8.1) and (8.2) in (8.6), we obtain

$$\dot{y} = \frac{k_1\left(v_{f1}\rho_1\left(1 - \frac{\rho_1}{\rho_{m1}}\right) - \beta U\right)}{\delta_1(k_2 - \rho_1)^2} + \frac{k_3\left(v_{f2}\rho_2\left(1 - \frac{\rho_2}{\rho_{m2}}\right) - \beta U\right)}{\delta_2(k_4 - \rho_2)^2} \tag{8.7}$$

which can be rewritten in the following form:

$$\dot{y} = F + G\beta \tag{8.8}$$

where

$$F = \left[-\frac{k_1 v_{f1} \rho_1}{\delta_1 (k_2 - \rho_1)^2} \left(1 - \frac{\rho_1}{\rho_{m1}}\right) + \frac{k_3}{\delta_2 (k_4 - \rho_2)^2} \left(\left(1 - \frac{\rho_2}{\rho_{m2}}\right) v_{f2} \rho_2 - U \right) \right] \quad (8.9)$$

$$G = \left(\frac{k_1}{\delta_1 (k_2 - \rho_1)^2} + \frac{k_3}{\delta_2 (k_4 - \rho_2)^2} \right) U \quad (8.9a)$$

Hence, a feedback linearization control law can be designed to cancel the nonlinearities and provide the desired error dynamics. The feedback control law can be written as

$$\beta = G^{-1}(-F + v) \quad (8.11)$$

which gives the closed-loop dynamics as

$$y = v \quad (8.12)$$

Simulation studies are performed to demonstrate the utilization of the feedback linearization technique presented. The test network, which consists of two alternative routes, is shown in Figure 8.6. We use a model with parametric uncertainties and partial user compliance to show the effectiveness of the controller even in the presence of such disturbances. A more comprehensive presentation of simulation results are given in [3].

The input function is assumed to be a sinusoidal function, which reaches a pre-defined peak value and then settles at a constant value for the rest of the simulation period. This function emulates the peak hour demand that reaches its maximum value at a certain time and then settles at a constant value when the peak period is over. In this specific simulation study, the peak period is assumed to be 1 hour.

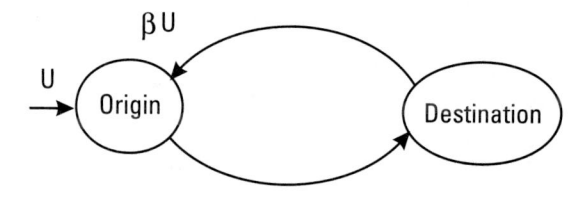

Figure 8.6 Sample network.

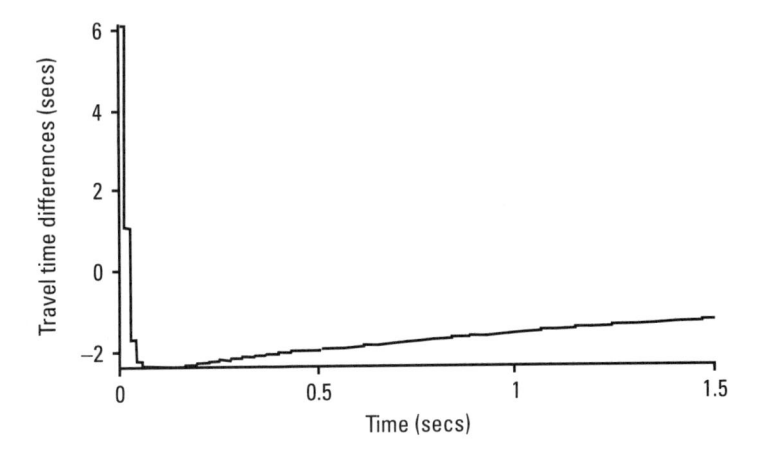

Figure 8.7(a) Differences in travel times (seconds).

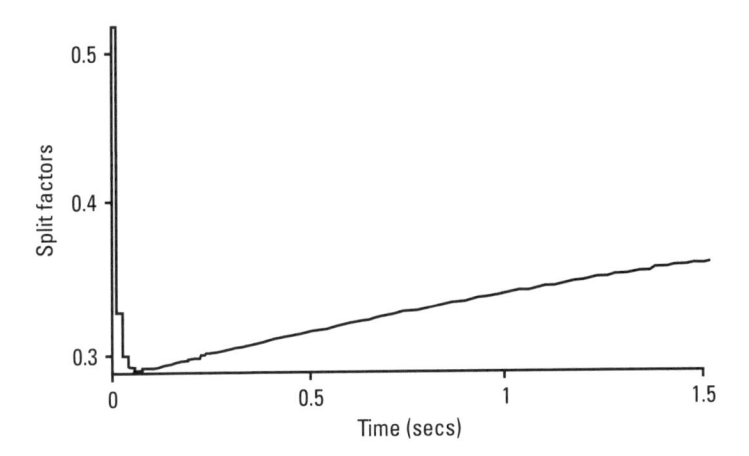

Figure 8.7(b) Split factors.

In this scenario, we assume both partial user compliance (80%) and the existence of parametric uncertainty in the model. As can be seen in Figure 8.7, the fluctuations of differences in travel are much higher than the previous scenarios, and it takes the controller a longer time to attain error convergence. However, even with partial user compliance and fairly large parametric uncertainties, the system stabilizes and the differences in travel times asymptotically converge at a desirable rate.

8.5 Summary

Feedback control design has been used effectively in various fields, such as electrical engineering (e.g., feedback amplifiers, robotics), mechanical engineering (e.g., machine control, noise control), industrial and systems engineering (e.g., process control). The technique can be highly effective in the emerging area of ITS and can provide real benefits, especially when compared to the traditional techniques being used.

This chapter discussed various traffic control problems that can be used for effective traffic management during incidents. Brief discussions on applications of real-time feedback control methods were provided. Using a sample problem, we showed that feedback control is a viable and attractive solution to the online dynamic traffic control/routing problem. To run the feedback control, we obtained desired states that the controller will track from the traffic sensors. In the control problem, the travel time to be tracked for each route was the travel time of the next route. The solution worked nicely for this simple problem under different demand and traffic conditions. We could try to generalize the concept for more complex networks, but more research needs to be done in this area to design for the states to be tracked by the controller.

This chapter should provide an incentive to further the development of feedback control design and application for traffic problems, especially DTA and DTR problems. Feedback control models, by making use of the sensor data and minimizing the need for expensive computational requirements of classical optimization techniques, provide a viable alternative in the context of ITS. Because most of the routing decisions are local and not networkwide, the solution for the type of DTR problem discussed in this chapter can be promising under incident conditions that do not require networkwide approaches.

Review Questions

1. Discuss the use of various types of sensors and actuation techniques for different traffic control problems.
2. Compare closed control versus open-loop control. Cite advantages and disadvantages of both.
3. Develop a simple macroscopic computer simulation program for two routes. Determine the split parameters using a simple splitting rule that will assign the incoming traffic to the two alternative routes. Discuss your results.

4. Prepare a list of online traffic control applications (field operation tests) that are currently being tested in this country and abroad. Choose a specific application and write a detailed review paper on it.

References

[1] Kachroo, P., and K. Ozbay, "Real Time Dynamic Traffic Routing Based on Fuzzy Feedback Control Methodology," *Transportation Research Record* 1556, 1996.

[2] Kachroo, P., et al., "System Dynamics and Feedback Control Formulations for Real Time Dynamic Traffic Routing With an Application Example" Intelligent Transportation Systems—Traffic Sensing and Management, *J. Mathematical and Computer Modeling (MCM)*, 1998, vol 27, no 9-11, pp. 27–49.

[3] Kachroo, P., and K. Ozbay, "Solution to the User Equilibrium Dynamic Traffic Routing Problem Using Feedback Linearization,", *Transportation Research: Part B.*, 1998, vol. 32:5, pp. 343–360.

[4] Kachroo, P., and K. Ozbay, "Feedback Control Solutions to Network Level User-Equilibrium Real-Time Dynamic Traffic Assignment Problems," *IEEE Southeastcon '97*, Blacksburg, VA, April 12–14, 1997.

[5] Kachroo, P., and K. Ozbay, "Sliding Mode for User Equilibrium Dynamic Traffic Routing Control," *IEEE Conf. Intelligent Transportation Systems ITSC'97*, Boston, MA.

[6] Kachroo, P., K. Ozbay, and A. Narayanan, "Investigating the Use of Kalman Filtering Approaches for Origin Destination Trip Table Estimation," *IEEE Southeastcon '97*, Blacksburg, VA, April 12–14, 1997.

[7] Wu, W., and P. Kachroo, "Dynamic Traffic Origin-Destination Estimation Using Kalman Filter: An Application to Beltway Network With VMS Control," *Proc. SPIE-Photonics East*, October 1997 (to be published).

[8] Gartner, N.H., and R.A. Reiss, "Congestion Control In Freeway Corridors: The IMIS System," *Flow Control of Congetsted Networks*, NAT, ASI seris, vol. F38, Springer Verlag, 1987.

[9] Friesz, T. L., et al., "Dynamic Network Traffic Assignment Considered as a Continuous Time Optimal Control Problem," *Operations Research* 37, 1989, pp. 893–901.

[10] Janson, B. N., "Dynamic Traffic Assignment for Urban Traffic Networks," Transportation Research, 25B, 1991, pp. 143–161.

[11] Jayakrishnan, R., A. Chen, and W. K. Tsai, "Freeway and Arterial Traffic Flow Simulation Analytically Embedded in Dynamic Assignment," Preprint #971342, TRB 76th Ann. Meeting, Washington, D.C., 1997.

[12] Ran, B., D. E. Boyce, and L. J. LeBlanc, "A New Class of Instantaneous Dynamics User-Optimal Traffic Assignment Models," *Operations Research* 41, 1995, pp. 128–142.

[13] Boyce, D. E., B. Ran, and L. J. LeBlanc, "Solving an Instantaneous Dynamic User-Optimal Route Choice Model," *Transportation Science*, Vol. 29, No. 2, 1995, pp. 192–202.

[14] Ozbay, K., and P. Kachroo, "Review Technologies and State of the Art Algorithms for Ramp Metering," *Proc. SPIE-Photonics East*, Pittsburg, PA, Oct. 1997.

[15] Wattleworth, J. A., *System Demand Capacity Analysis on the Inbound Gulf Freeway*, Texas Transportation Institute Report 24-8, 1964.

[16] Drew, D. R., *Gap Acceptance Characteristics for Ramp Freeway Surveillance and Control*, Texas Transportation Institute Report 24-12, 1965.

[17] Pinnell, C., et al., *Inbound Gulf Freeway Ramp Control Study I*, Texas Transportation Institute Report 24-10, 1965.

[18] Pinnell, C., et al., *Inbound Gulf Freeway Ramp Control Study II*, Texas Transportation Institute Report 24-10, 1965.

[19] Institute of Transportation Engineers, *ITE Traffic Control Systems Handbook*, 1996, pp 4–14 to 4–40.

[20] Yuan, L. S., and J. B. Kreer, "Adjustment of Freeway Ramp Metering Rates to Balance Entrance Ramp Queues," *Transportation Research*, Vol. 5, 1971, pp. 127–133.

[21] Wattleworth, J. A., and D. S. Berry, "Peak Period Control of a Freeway System—Some Theoretical Investigations," *Highway Research Record*, Vol. 89, pp. 1–25.

[22] May, A. D., "Optimization Techniques Applied to Improving Freeway Operations," *Transportation Research Record* 495, pp. 75–91.

[23] Hellinga, B., and M. Van Aerde, "Examining the Potential of Using Ramp Metering as a Component of an ATMS," *Transportation Research Record* 1494, 1995, pp. 75–83.

[24] Van Aerde, M., "Evaluation of Ramp Metering System Design," Working Paper, Department of Civil Engineering, Queen's University, Kingston, Ontario, Canada, 1997.

[25] Papageorgiou M., H. S. Habib, and J. M. Blosseville, "ALINEA: A Local Feedback Control Law for On-Ramp Metering," *Transportation Research Record* 1320, 1991, pp. 58–64.

[26] Chen, L. L., A. D. May, and D. M. Auslander, "Freeway Ramp Control Using Fuzzy Set Theory for Inexact Reasoning," *Transportation Research A*, Vol. 24A, No. 1, 1990, pp. 15–25.

[27] Goldstein, N. B., and K. S. P. Kumar, A Decentralized Control Strategy for Freeway Regulation," *Transportation Research-B*, Vol. 16B, No. 4, 1982, pp. 279–290.

[28] Zhang, H., S. G. Ritchie, and Z. Lo, "A Local Neural Network Controller for Freeway Ramp Metering," *IFAC Transportation Systems*, Tianjin, PRC, 1994, pp. 655–658.

[29] Masher, D. P., et al., *Guidelines for Design and Operation of Ramp Control Systems*, Stanford Research Institute, Menid Park, CA, 1975.

[30] Hadj-Salem, H., et al., *ALINEA: Un Outil de Regulation d'Acces Isole sur Autoroute*, Rapport INRETS 80, Arcueil, France, 1988.

[31] La Pera, R., and R. Nenzi, "TANA—An Operating Surveillance System for Highway Traffic Control," *Proc. IEEE*, Vol. 61, 1973, pp. 542–556.

[32] Helleland, N., W. Joeppen, and P. Reichelt, "Die Rampendosierung an der A5 Bonn/Siegburg der BAB 3 in Richtung Koln," *Strassenverkehrstechnik*, Vol. 22, 1978, pp. 44–51.

[33] "NN: Ramp Metering in Auckland," *Traffic Engineering and Control*, Vol. 24, 1983, pp. 552–553.

[34] Owens, D., and M. J. Schofield, "Access Control on the M6 Motorway: Evaluation of Britain's First Ramp Metering Scheme," *Traffic Engineering and Control*, Vol. 29, 1988, pp. 616–623.

[35] Buijn, H., and F. Middelham, "Ramp Metering Control in the Netherlands," *Proc., 3rd IEE Internat. Conf. Road Traffic Control*, United Kingdom, 1990, pp. 199–203.

[36] Gordon, R. L., et. al., *Traffic Control Systems Handbook*, FHWA-SA-95-032.

[37] Lighthill, M. J., and G. B. Whitham, "On Kinematic Waves II. A Theory of Traffic Flow on Long Crowded Roads," *Proc. Royal Society of London*, Series A 229, 1955, pp. 317–345.

[38] Kachroo, P., "System Dynamics and Feedback Problem Formulations for Real-Time Ramp Metering Control," *1998 Transportation Research Board Annual Meeting*, Washington, D.C.

[39] Ball, J. A., et al., "Simulation Study of a Hybrid System Formulation With H-Infinity Feedback Control for Signalized Intersection," *IEEE Conf. Intelligent Transportation Systems ITSC'97*.

[40] Papageorgiou, M., and A. Messmer, "Dynamic Network Traffic Assignment and Route Guidance via Feedback Regulation," *Transportation Research Board*, Washington, D.C., Jan. 1991.

Selected Bibliography

Ball, J. A., P. Kachroo, and T. Yu, "Real-Time Demand-Driven Signalized Intersections," *Proc. SPIE-Photonics East*, Oct. 1997 (to be published).

Messmer, A., and M. Papageorgiou, "Optimal Freeway Network Control via Route Recommendation," *Vehicle Navigation & Information Systems Conf. Proc.*, 1994.

Messmer, A., and M. Papageorgiou, "Route Diversion Control in Motorway Networks via Nonlinear Optimization," *IEEE Trans. Control Systems Tech.*, Vol. 3, No. 1, March 1995.

Zhang, H. And S.G. Ritchie, "An Interpreted Traffic Responsive Ramp Control Strategy via Nonlinear State Feedback," Preprint #950903, *Transportation Research Board 74th Annual Meeting*, January 22-28, 1995, Washington, D.C.

9

Conclusions and Future Research

9.1 Conclusions

Traffic incidents are the major cause of non-recurrent congestion and delay in freeway systems. Automated incident management plays a vital role in managing and expediting the incident response and clearance processes and reducing congestion and delay. In this book, different steps of incident management process have been described. More important, most of the individual steps have been illustrated using the WAIMSS example developed at Virginia Polytechnic Institute and State University. WAIMSS was developed with the goal of providing incident management personnel with appropriate strategies to manage traffic incidents and to execute the steps required for implementation of those strategies.

The concept of computerized incident management support process was described using the WAIMSS architecture. WAIMSS is a GIS expert system that combines the powerful spatial data-handling capabilities of a GIS with the rule-based logic of an expert system in a fully integrated Expert-GIS framework to provide interactive content and group process support for incident management operations. Linking an expert system with a GIS enables both the expert system and the GIS to perform new tasks, and opens the way for more complex spatial analysis and more flexible querying of GIS databases. That is an important requirement for effective incident management support of real-time incident management operations.

WAIMSS does not have a module that detects and verifies incidents. However, this important aspect of incident management is also discussed in the book and a brief review of some of the existing incident detection techniques has also been provided. The WAIMSS software has the following functional modules, each reflecting a different stage of the incident management decision making process.

9.1.1 Incident Input

WAIMSS blackboard architecture allows users of the system to input characteristics of the incident irrespective of their geographical locations as long as they are connected to the server that runs WAIMSS. That has been made possible through the implementation of the WAIMSS using a client/server architecture. Users are defined as clients. The database and the decision-making capabilities of the system reside on the server. The GIS component has attractive features such as graphical input/output, map displays, and address matching. WAIMSS also features additional GUI tool kits built using open-interface elements for easy input/output.

9.1.2 Duration Estimation and Delay Prediction

WAIMSS has the ability to estimate the duration of a specific incident based on its characteristics using decision/prediction trees. These trees were constructed as a result of rigorous statistical analysis of the data collected from over 6,000 incidents in Fairfax County, Virginia.

Incident delay information is also predicted based on a variety of factors, including network characteristics, lanes blocked, travel demand, and estimated incident duration. A simple deterministic queuing model is used to predict the delay caused by each incident using the incident duration information obtained in the incident duration estimation stage.

9.1.3 Response Plan Development

Development of an efficient response plan presents a great challenge because of the differences in the needs and capabilities of different areas. The response module of WAIMSS suggests suitable measures to efficiently clear an incident using the same incident data employed to develop duration estimation models. The module also provides the user with recommendations as to which agencies and personnel to contact and which and how many response units and equipment to dispatch. The response information has been programmed in the form of rules and is part of the WAIMSS knowledge base. It is, however, important to note that the response recommendations of WAIMSS are based on the state of the practice that existed in northern Virginia at the time of the system development. The major challenge in developing any new operational response model similar to WAIMSS will be to update the WAIMSS knowledge base to reflect the differences in the state of the practice.

9.1.4 Traffic Diversion and Control

Traffic diversion and control is one of the most active research areas of incident management. The WAIMSS decision support module for traffic diversion proposes a unique approach. It has the capability to obtain predefined diversion plans by two methods:

- From a database of diversion routes currently used by VDOT in the northern Virginia area;
- By dynamically generating diversion plans using a rule-based system developed at Virginia Tech.

Moreover, the network generator module of WAIMSS eliminates links that are infeasible for diversion based on real-time conditions such as high traffic congestion, special events, and inclement weather. It can then generate new dynamic diversion routes using only the feasible links.

In addition to the above-mentioned modules of WAIMSS, two separate issues were discussed in Chapters 4 and 8. Chapter 4 presented a summary of the incident detection issues. In addition, several operational field tests conducted to evaluate the use of new communication technologies were described. Examples such as cellular phones and electronic toll tags for incident management are but two of those new technologies. Chapter 8 introduced traffic control using feedback control techniques. To give the user a general idea about the application of this new concept, a sample problem was studied.

9.2 Future Research

Future research should focus on the following areas.

9.2.1 Incident Detection

The focus of attention should be on using multiple data sources such as loop detectors, cellular phones, and electronic toll tags. That would be possible only through the use of efficient and reliable data fusion algorithms that are field tested. A growing area of research is incident detection on surface streets. Interrupted traffic flow due to traffic signals makes reliable incident detection on the arterials a challenging problem. However, it is clear that effective arterial incident detection will drastically improve the efficiency of incident management, and thus the quality of traffic over the entire transportation network, by reducing the time to clear the incidents.

9.2.2 Validation and Elaboration of Duration Prediction

The existing duration prediction models similar to the prediction/decision trees can be further elaborated, validated, and expanded with more new online data. Focus should be on severe incidents with large duration and for which no sufficient data are available. More reliable and accurate duration estimation models will improve incident management by providing traffic operators with valuable information to develop traffic management strategies.

9.2.3 Real-World Implementation of Duration and Delay Models

Although the duration prediction study results of reviewed projects, including WAIMSS, are comparable, the transferability of the models to other areas is a challenge that deserves serious attention. Nevertheless, different network characteristics, traffic demand levels, and differing practices and procedures of response organizations make the incident clearance process area specific. Thus, model validation and tuning are necessary for different implementation areas. To implement the duration module in any given area of the country, the following steps need to be considered:

- *Detailed data-needs study.* Based on the previous study and developed decision/prediction trees, a set of survey forms/data item lists should be determined for data collection of different types.
- *Data collection.* Data collection can take different approaches. An example of one approach would include recording incident clearance process/duration using video display at the traffic control center. At the same time, incident response personnel from different agencies can record the clearance process using survey forms or PC-based software.
- *Duration model validation and enhancement.* Collected data can be used to validate developed models and enhancements made accordingly.

9.2.4 Advanced Traffic Control Algorithms

The use of real-time traffic control algorithms is essential to alleviate congestion caused by major accidents. Feedback-based control models currently researched hold great potential for the development of such real-time control models.

9.2.5 Evaluation of Existing Incident Management Programs

Incident management programs have to be evaluated to assess their benefits and costs. This is becoming an important issue among traffic engineers and policy makers. New incident management practices that maximize benefits while minimizing costs also can be developed using the results of those studies. The quantification of benefits and costs of incident management programs require collection of before and after data. Because many incident programs have been in operation for several years, getting before data might be a problem. In such situations, simulation programs developed specifically for simulating incident management operations can be used. However, well-tested and validated simulation programs capable of handling different incident management procedures are not available to traffic engineers. That is another challenging research area for the future.

About the Authors

Dr. Kaan Ozbay, received his B.S. in Civil Engineering in 1988 from Bogazici University, Istanbul, Turkey, his M.S. in 1991 in Civil Engineering (Transportation) from Virginia Tech, and his Ph.D. in Civil Engineering (Transportation) in 1996. Dr. Ozbay's research interest in transportation covers advanced technology applications in ITS, incident management, development of real-time control techniques for traffic, application of artificial intelligence and operations research techniques in network optimization, and development of simulation models for automated highway systems. Dr. Ozbay joined Rutgers University Department of Civil and Environmental Engineering as an assistant professor in September, 1996. He is currently the associate director of the newly established Rutgers "Center for Advanced Infrastructure and Transportation (CAIT)" sponsored by the Federal Highway Administration's University Transportation Centers (UTC) program. As the associate director of CAIT, he is mainly responsible of the Center's Intelligent Transportation Systems (ITS) research projects and other related educational and administrative activities.

In addition to this book, he and Dr. Pushkin Kachroo of Virginia Tech are co-authors of a book titled *Feedback Control Theory for Dynamic Traffic Assignment*. Dr. Ozbay has published more than 30 papers in scholarly journals and conference proceedings. Recently, he was the chairman of several ITS sessions of the SPIE conferences in 1997 and 1998 and 2 sessions at the IEEE SMC conference held at San Diego in 1998. He is also the co-editor of the 1997 SPIE ITS session proceedings published by SPIE.

He is currently the principal investigator of two projects namely, "Cost of Transportation in New Jersey" sponsored by the New Jersey Department of

Transportation (NJDOT) and the University Transportation Research Center at CUNY and "Pavement Deterioration Models" sponsored by NJDOT. He recently completed several projects, namely, "Technical Support to Incident Management" sponsored by the Virginia Tech Center for Transportation Research and was the co-principal investigator of projects titled "Logical and Physical Simulation of Automated Dedicated Bus Lanes" sponsored by the University Transportation Research center at CUNY and "Alternate Bus Routing System Evaluation Project" funded by the New Jersey Highway Authority.

Between 1993 and 1996, Dr. Ozbay was a research associate and senior research associate at the Virginia Tech Center for Transportation Research. At the Center, he was the co-principal investigator and project manager of several ITS projects including the "Wide-Area Incident Management Expert System" project sponsored by FHWA/VDOT and the co-principal investigator of the "Dynamic Network Optimization," and "Automated Incident Management," projects sponsored by the FHWA ITS Research Center of Excellence.

Dr. Pushkin Kachroo is an Assistant Professor at the Bradley Department of Electrical and Computer Engineering at Virginia Tech. He obtained his P.E. license in Electrical Engineering in 1995 from the State of Ohio, Ph.D. from University of California at Berkeley in 1993, M.S. from Rice University, Houston in 1990, and his B.Tech from the Indian Institute of Technology, Bombay in 1998. Apart from this book, he is also the co-author with Dr. Kaan Ozbay of the book *Feedback Control Theory for Dynamic Traffic Assignment*. He has been the chairman of the robotics or ITS sessions of the SPIE conference in 1996, 1997, and 1998. He is the editor of the proceedings of those sessions published by SPIE. He has more than fifty papers published in journals and conference proceedings. His teaching and research are in the areas which include nonlinear control theory and applications to transportation and communication systems, electronics, and microcontroller based embedded systems. He has been a research engineer at the robotics R&D laboratory at Lincoln Electric Company for two years and a research scientist at the Center for Transportation Research at Virginia Tech for three years.

Index

ACC. *See* Alcohol consumption per capita; Chicago Area Expressway Accidents Annual Report
Accident notification time, 63
Accident with personal injuries, 88, 105, 109–10, 115, 158–59
Accidents involving property damage, 88, 104, 108, 114, 148
Acoustic detection, 66
ACPD. *See* Accidents involving property damage
ACPI. *See* Accident with personal injuries
Actuation, 216–17
ADVANCE project, 74, 87–89
Aerial surveillance, 12
Agency coordination, 109–10, 133, 142, 150–51
AI-Client, 57
AID. *See* Automated incident detection
Airports, 193
AI-Server, 57
Alcohol consumption per capita, 63
Alternative routing, 35
Ambulance involvement, 109–11, 123, 157, 161
American Traveling Association, 17
AML. *See* Arc Macro Language
Analysis of variance, 88, 99, 107–9
Anderson-Darling test, 121

ANN. *See* Artificial neural network
ANOVA. *See* Analysis of variance
ANT. *See* Accident notification time
Antecedent, rule, 189–90, 200–201
API. *See* Application programming interface
Application programming interface, 56–57
Arc/Info, 36, 50, 52–55, 57, 62, 153, 177–78
Arc Macro Language, 52–55, 57
Arrowboard usage, 157
Arterial incident, 192
Artificial intelligence, 25–28, 73
Artificial neural network, 73
Asynchronous events, 24
Automated incident detection, 62–63, 66, 68
point-based algorithms, 69–73
spatial measurement-based algorithms, 70, 74
Automatic traffic control, 216, 231

Baltimore beltway, 168–69
Bayesian algorithm, 71
Binary split prediction model, 99
Blackboard architecture, 25–26, 28–31, 43
control component, 30
data structure, 29–30
knowledge source, 29–30
wide-area system, 43–45, 232

Boolean logic, 189–90
Boy/Jansen traffic network study, 181
Bridges, 3, 109

C programming language, 45, 47–48, 50, 56–57, 177
CAD. *See* Computer-aided dispatch
California algorithm, 70–72
California Department of Transportation, 143
California Freeway Service Patrol study, 65
California Highway Information Network, 138
California Highway Patrol, 65, 138
California truck accident study, 85–86
Caltrans incident management program, 137–40
Cambridge Systematics study, 7–8
CAMEO, 35
Capital spending, 3
Cargo spill, 149
CART. *See* Classification/regression tree
Catastrophe theory, 72
CB radio. *See* Citizens band radio
CCD. *See* Charged coupled device
CCTV. *See* Closed-circuit television
Cellular phone notification, 11, 65, 77–78, 233
Certainty factor, network generator, 197
Changeable message sign, 13, 31–32, 49
Charged coupled device, 66
Chicago, 6, 172
Chicago Area Expressway Accidents Annual Report, 86
CHIN. *See* California Highway Information Network
Chi-square test, 121–22
CHP. *See* California Highway Patrol
Citizens band radio, 62
Classification/regression tree, 99
Clearance. *See* Incident clearance
Client-centric architecture, 52
Closed-circuit television, 11, 31, 62, 77
Closed-loop traffic control, 212–13, 217, 219, 224
CMS. *See* Changeable message sign
Comfort's system architecture, 25–28

Command-level integration, 56–57
Comparative AID algorithm, 70
Compound link elimination, 201
Computer-aided dispatch, 65
Computer-based decision support tool, 137
Computer-based response plan, 153
Conditional probability factor, 84–85
Conflict resolution, link elimination, 200
Congestion. *See* Highway congestion
Connect-file, 53
Connecticut, 140
Construction, road, 13, 192
Corridor-wide diversion strategy, 183
Coverage (storage unit), 54
Critical elimination rule, 198–99
Cumulative weight function, 200

Data accuracy/analysis, 25, 62–63, 67–68
Data collection, 234
Data-level integration, 56
Data-needs study, 234
Decision support module, 46–47
Decision tree logic, 70–72, 98, 129
    Virginia case study, 98–99, 112–20,232
Definitely don't eliminate link rule, 195
Definitely eliminate link rule, 194, 198–99
Delaware, 140
Delay
    dispatch time, 135, 146
    economic effects, 3–4
    See also Incident delay prediction; Incident duration
Department of Public Works, 140, 151
Department of Transportation, 12–13, 44, 50, 66, 140, 150
Detection. *See* Incident detection and verification
Deterministic queuing, 48, 83, 125–27
Detroit, 62
Disablement incident, 6–7, 12, 111, 116–17, 149
Dispatch delay time, 135, 146
Diversion. *See* Traffic diversion
Don't eliminate link rule, 195
Doppler radar detection, 67
DOT. *See* Department of Transportation

Index

DPW. *See* Department of Public Works
DTR/DTA. *See* Dynamic traffic routing/
    assignment
Duration. *See* Incident duration
Dynamic link elimination. *See* Link
    elimination
Dynamic network information, 53, 170
Dynamic traffic routing/assignment,
    167, 212–18
    feedback control, 220–22

Electronic toll and traffic management, 76
Eliminate link rule, 194, 199
Emergency medical services, 150
Emergency response unit, 145
Environmental factors, 109, 111
Environmental Protection Agency, 13
Environmental Systems Research
    Institute, 54
EPA. *See* Environmental Protection Agency
ERU. *See* Emergency response unit
ESRI. *See* Environmental Systems Research
    Institute
ETTM. *See* Electronic toll and traffic
    management
Evaluation, detection system, 74–75
Event-scanning traffic simulation, 168–69
Expert knowledge, 155
Expert system, 23–24, 30–32
    Caltrans, 137–40
    integration with GIS, 45–47, 54
    real-time knowledge-based, 24–25, 32
    traffic diversion model, 169–70
    See also Network generator
External interface, 25

Fairfax County agencies, 93, 95–97, 154
False alarm rate, 76–77
Fatalities, 63, 148, 150
Federal Highway Administration, 4, 22, 33,
    70, 89, 170
Feedback control, 177, 211, 217–22, 235
    example, 222–25
Feedback linearization control law, 224
Feedback loop, 23
FHWA. *See* Federal Highway
    Administration

Filtering algorithm, 72
Fire incident, 106, 111–12, 123,
    149–51, 156
FRED. *See* Freeway real-time expert-system
    demonstration
Freeway capacity, 7–8
Freeway incident, 103, 127
    prediction model, 89–90
Freeway real-time expert-system
    demonstration, 31–32
Freeway Service Patrol, 10, 12, 65,
    77, 143–44
Freight services demand, 4
FSP. *See* Freeway Service Patrol
Fuel, wasted, 3–4

GDSS. *See* Group decision support system
Geographical Information Systems, 30,
    33–36, 54–56, 137, 153, 177–78,
    182, 231–32
Geometric characteristics, link, 193
GIS. *See* Geographical Information Systems
Graphical user interface, 50, 56, 232
Group decision support system, 43–45
Group process support, 43
Group support system, 43
GSS. *See* Group support system
GUI. *See* Graphical user interface

HA. *See* Highway authority
Hall traffic diversion study, 168
HAR. *See* Highway advisory radio
Hazardous material accident, 7–8, 13,
    16–17, 89, 102, 112, 123, 149–51
HAZMAT. *See* Hazardous material accident
HEARSAY-II, 28
Highway advisory radio, 49, 138, 142,
    151, 177
Highway authority, 140
Highway capacity growth, 3
Highway congestion, 1–4
    impact of incidents, 4–6
    incident types, 1, 4, 6–9, 102–7, 146
Historical data, 155
Hospitals, 193
Houston GIS project, 34–35
Hughes Aircraft study, 67

Human-in-the-loop control, 216
Hybrid incident management system, 30–31, 182

IAC. *See* Interapplication communication
Ice clearance, 193–94
Illinois Department of Transportation, 86
Illinois traffic network, 180–81
IMPACT prediction model, 89–90
Impedance function criterion, 205–6
Incident characterization, 147–49
Incident clearance, 8–9, 12–13, 16
   process tasks, 151–52
   time for, defined, 135
      Virginia study definition, 93
Incident database query, 36
Incident delay prediction, 83, 125–28, 232
   calculation module, 46, 48–49
   deterministic queuing, 125–27
   See also Incident duration
Incident detection and identification time, 145
Incident detection and verification, 9–11, 12–13, 24, 46, 61
   algorithmic issues, 69–74
   conclusions, 233–34
   defined, 62
   detection time, 63–66, 134
   operational field tests, 76–78
   surveillance issues, 66–69
   traffic data analysis, 62–63
   traffic surveillance, 62
   verification issues, 74–75
Incident duration, 64–66, 83–84
   detection time and, 64–66
   estimation module, 46–48
   estimation models, 84–91, 232, 234
   incident type and, 6–10, 102–7
   See also Virginia case study
Incident impact area
   conflict resolution weighting, 200
   defined, 183–84
   diversion volume estimation, 186–88
   dynamic link elimination, 188–89, 194–95
   estimation, 183–84
   knowledge representation, 184–86
   link elimination, 189–94

link elimination decision making, 195–99
link elimination rules, 196–97, 200–1
Incident management, 9–12
   agencies involved, 12–14
   major goals, 13
   major steps, 14–15
   problems, 15–17
   program evaluation, 235
      support tools, 17
Incident management systems, 21–22
   blackboard architecture, 28–31
   expert systems, 31–33
   geographical information systems, 33–36
   implementation frameworks, 22–31
   See also Wide-area incident management support system
Incident recovery, 9, 12, 46
Incident response, 11–13, 29, 35, 133–35
   activity time, 146
   Caltrans system, 137–40
   four components, 134–35
   I-95 corridor coalition, 140–44
   northern Virginia system, 144–45
   research, 145–46
   tools, 135–37
   two stages, 133–34
Incident response plan, 15, 45–47, 53
   agency notification, 150–51
   clearance process, 151–52
   computer-based, 153
   incident characterization, 147–49
      module, 49–50
   service identification, 149–50
   steps, 146–47
   Virginia case study, 154–62, 232–33
Incident severity, 109–10, 184–85, 190–91
Incident site management, 149
Incident type, 6–10, 102–7, 146
Income per capita, 63
Inductive loop, 66–67
Information dissemination, 13, 129, 134, 142, 150–51, 172
Infrared technology, 66
Injuries data, 148
Intelligent transportation systems, 34, 63, 66, 211
   dynamic traffic routing, 212–18

Interactive interface, 49–50
Interapplication communication, 52, 55, 57
Inter-freeway diversion strategy, 183
Interstate 880 study, 65, 77–78, 90–91, 127–28
Interstate 95, 11, 13
Interstate 95 Corridor Coalition, 140–44
IPC. *See* Income per capita
ITS. *See* Intelligent transportation systems

Jurisdictional factors, 192
Just-in-time manufacturing, 4

Kappa-PC, 31
KBES. *See* Real-time knowledge-based expert system
Knowledge acquisition, 26
Knowledge-based decision support, 24–25
Knowledge-based expert system, 25, 32
    See also Network generator
Knowledge representation, 26, 28
Knowledge source, 29–30
Knowledge utilization, 28

Land use type, 2, 109
Lane blockage data, 148
Lane closure, 106, 112, 116–17, 124–25
Linear regression prediction model, 97–98
Link capacity, 193
Link elimination, 188–90
    conflict resolution, 200
    decision making, 195–99, 233
    factors influencing, 190–94
    probabilistic knowledge, 189–90
    rule antecedents, 200–201
    rule base, 194–95
    rule classification, 201
    rule structure, 196–97
    simple and compound, 201
Link location, 193
Link significance, 179
Link type, 193–94
Local agency, 44
Location factors, 109, 111, 148, 193
Location-specific rules, 185
Log-linear modeling, 85–86
Log-logistic function, 84–85, 98
Log-normal distribution model, 85–86, 89–90, 98, 121
Long Island freeway network, 170
Loop detector, 11, 62, 66–67, 77, 233
Los Angeles, 5–6, 169
Lumped parameter setting, 220–21

MA. *See* Motorist assist
McMaster algorithm, 72–73
Macroscopic traffic control, 211, 218–22
Magnetometer, 66
Maine, 140
Maintenance factors, 13, 192
Major axis, impact area, 184–85
Management. *See* Incident management; Incident management systems
Maryland, 140
Massachusetts, 31, 140
Mean vehicle speed, 63
Measure of effectiveness, 74–75, 145
Media notification, 49, 151
Medical evacuation, 13, 151
MEDVAC. *See* Medical evacuation
Metros, traffic diversion and, 193
Microscopic traffic control, 211
Military involvement, 151
Minor axis, impact area, 185
MOE. *See* Measure of effectiveness
Motorist assist, 88
Motorist compliance, 171–72, 184, 205–6
Multiagency network, 22, 42
Multiple-point diversion, 206–7
MVS. *See* Mean vehicle speed
MYCIN system, 190, 195, 200

National Incident Management Coalition, 135
NEA. *See* Network element abstraction
NEE. *See* Network element extraction
Neighborhood factor, 192
Network aggregation model, 178–82
Network connectivity, 207
Network element abstraction, 179–80
Network element extraction, 179–80
Network generator, 175–77
    certainty factor use, 197
    compared to network aggregation, 181–82

diversion strategies, 182–83
diversion volume estimation, 186–88
function and theory, 177–78
incident impact area, 183–86
link elimination, 189, 195, 197
route generation, 201–5
Neural network prediciton model, 98
New Hampshire, 140
New Jersey, 13, 140, 143, 168
New Jersey Institute of Technology, 76
New York City, 5–6
New York State, 13, 143
Nexpert-Object, 31, 47, 50, 54–57, 177–78
Nexpert rule, 52, 55
NJIT. *See* New Jersey Institute of Technology
No-diversion strategy, 183
Nonlinear H8 traffic control, 222
Nonlinear numerical modeling, 219
Nonrecurrent congestion, 1, 4
Non-vehicle incident, 146
Northwest Central Dispatch, 87
Northwestern University ADVANCE Project, 74, 87–89
Northwestern University incident clearance study, 86–87
NOVA, 137
NP. *See* Number of police cars dispatched
Number of fire trucks dispatched, 88
Number of police cars dispatched, 88
NWCD. *See* Northwest Central Dispatch

ODE. *See* Ordinary differential equation
OD pair. *See* Origin-destination pair
Office hours, 192
ONC. *See* Open Network Computing
Online traffic control, 211
  dynamic routing/assignment, 212–13
  feedback control, 218–25
  ramp metering, 216–18
  signalized intersection control, 218
  traditional techniques, 213–16
  traffic speed control, 218
Open Interface Elements, 50, 56
Open-loop traffic routing, 212–13, 219
Open Network Computing, 57
Operational factors, 108–11

Operational field test, 76–78
Optimal resource allocation, 135–36, 154–62
Optimization-based traffic control, 212, 214–15, 219–20
Orange County, California, 31, 137–40
Ordinary differential equation, 220–21
Origin-destination pair, 179, 207
Overseas part suppliers, 4

Partial differential equation, 220–21
PASSER-II software, 35
Pattern recognition-based algorithm, 70
PDE. *See* Partial differential equation
Peak-hour travel, 1
Pennsylvania, 140, 143
Performance, incident management, 24
Personal computer platform, 36
Personal injury incident, 88, 105, 109–10, 115, 158–59
PID control. *See* Proportional-integral-derivative control
Point-based algorithm, 69–73
Point diversion, 183, 206–7, 215, 218–22
Poisson equation, 63–64
Police involvement, 10–12, 44, 108, 110, 123, 149–50, 158
Population growth, 2
Post-incident traffic control, 33
Predefined diversion plan, 201
Pre-selection of routes, 170
Prioritized action list, 53
Probabalistic functions, route prioritization, 205
Probabilistic knowledge, 189–90, 196–97
Problem content support, 43
Process list, 54
Production rules, 184–86, 189–90, 195
Property damage incident, 88, 104, 108, 114, 148
Proportional-integral-derivative control, 220
Pulsed radar technology, 67

Queuing delay approach, 8–10

Radio broadcast notification, 49, 142
Ramp meter, 31, 216–18
Real-Time Adaptive Control Strategies, 167

Real-time expert system
    Caltrans, 137–40
    knowledge-based decision
        support, 24–25
    knowledge-based expert system, 25, 32
    models, duration estimation, 83–84
Real-time performance, 68
Real-time traffic diversion, 168–72
Recovery. *See* Incident recovery
Recurrent congestion, 1, 4
Reliability, detector technology, 67–68
Remote Procedure Call, 57
Resource allocation/assesssment, 16, 35, 54
    optimal, 135–36
    Virginia case study, 154–62
Response. *See* Incident response; Incident
        response plan
Rhode Island, 140
Ritchie-Prosser decision support
        architecture, 24–26
Road damage repair, 150
Road database query, 36
Road hazard incident, 104, 107
Roadway type, 109
Route generation, 49, 188, 201, 204–5
Route prioritization, 205–6
Route system, 50
RPC. *See* Remote Procedure Call
RT-TRACS. *See* Real-Time Adaptive
        Control Strategies
Rubbernecking, 7
Rule base, link elimination, 45, 194–99,
        202–4

San Francisco, 6
SAS. *See* Statistical analysis software
School hours, 192
SCOT. *See* Simulation of corridor traffic
Seattle duration estimation study, 84–85
Sensing technology, 216
Sensor interface, 25
Sequential-gradient-restoration
        algorithm, 219
Service identification, 149–50
Service level D/E, 3
Service patrol, 10, 12, 65, 77, 143–44
Severe incident, 88–89, 110–11, 124

Severity, incident, 109–10, 184–85, 190–91
Shortest-path algorithm, 49, 201,
        204–5, 207
SI. *See* Severe incident
Siegfried-Vaidya prototype, 34–36
Signalized intersection control, 167,
        211, 218
Simple link elimination, 201
Simple Macro Language, 36
Simulation-based stochastic model, 83–84,
        127–28
Simulation of corridor traffic, 168
Single-input, single-output system, 220
SISO system. *See* Single-input, single-output
        system
SML. *See* Simple Macro Language
SND. *See* Standard Normal Deviation
Snow clearance, 193–94
Software implementation, WAIMSS, 50–52
    application development, 54–56
    architecture, 52–54
    command-level integration, 56–57
    data-level integration, 56
Space-discretized model, 221–22
Spatial analysis, 34, 43, 45, 54
Spatial management, incident response, 145
Spatial measurement-based detection, 70, 74
Special events, 192
Speed limit, 193
Sperry Rand Corporation, 168
Split factor, traffic diversion, 215–16,
        223, 225
SQL. *See* Structured Query Language
SSID. *See* Surface street incident detection
Standard Normal Deviation, 70–71
Starting point, traffic diversion, 186
Static network information, 53
Static traffic assignment, 179
Static traffic diversion, 212
Statistical AID algorithm, 70–72
Statistical analysis software, 95
Structured Query Language, 155
Suburban highway congestion, 1–2
Suffolk traffic management system, 167
Surface street incident detection, 74
Surveillance. *See* Traffic surveillance
Survey methodology, 91–98, 154

System optimal, 212, 214–15, 219–20
TA. *See* Turnpike authority
TASAS. *See* Traffic Accident Surveillance and Analysis System
TCS. *See* Traffic control system
Television notification, 49
Temporal data analysis, 43
Termination point, traffic diversion, 186
Texas A&M study, 172
Texas Department of Transportation, 216–17
Texas Transportation Institute, 70, 205–6
Theoretical algorithm, 72–73
Time of detection, 148
Time of occurrence, 148
Time-sequential incident clearance prediction model, 87
Time series algorithm, 72
Time to normal flow, 9, 125–27
TIST. *See* Total incident service time
TMC. *See* Traffic management center
TMO, 152
TNF. *See* Time to normal flow
TOC. *See* Traffic operations center
Torino, Italy, 171–72
Total incident service time, 145
Tow truck clearance, 12, 149, 151
Traffic Accident Surveillance and Analysis System, 85
Traffic control system, 34
  See also Online traffic control; Traffic management
Traffic data analysis, 62–63
Traffic diversion, 123, 149–50
  control/routing module, 177
  initiation module, 174–75
  models for, 168–72
  network aggregation models, 178–82
  network generator, 177–78, 182–83
  rate system, 50
  solution approach, 165–67
  strategy planning module, 175–77
  volume estimation, 186–88
  WAIMSS architecture, 172–77, 233
Traffic flow, 69, 72–73

Traffic management, 17, 134, 138, 142, 149–50
Traffic management center, 11, 22, 44, 45, 52, 54, 150, 154, 166
Traffic operations center, 12–13, 22, 24–25, 53, 62, 66, 135
  incident report, 146–47
Traffic speed control, 218
Traffic stop/arrest, 87
Traffic surveillance, 62, 66–69
Traffic volume, defined, 191
Traffic volume-to-capacity ratio, 3
TRAFLO, 33, 170
TRANSCOM, 13, 76–77
Transguide system, 216–217
Travel lane incident, 106, 112, 116–17, 124–25
Travel time, response team, 135, 146
Truck driver travel preferences, 171–72
Truck incident, 16–17, 105, 108, 111, 116
  California study, 85–86
Trucking industry effects, 3
Truth maintenance, 24
TSA. *See* Traffic stop/arrest
TTI. *See* Texas Transportation Institute
Tunnels, 109
Turnpike authority, 140

Ultrasonic detection, 66
University of California, Irvine, 138
Urban highway congestion, 1–6
User-equilibrium assignment module, 31

Variable message sign, 13, 129, 134, 142, 149, 151, 166, 177
Vehicle incident, 146
Vehicle ownership statistics, 2
Vehicles involved data, 148
Vehicles miles traveled, 63
Vehicle usage habits, 2
Verification, incident, 61
  See also Incident detection and verification

Vermont, 140
Virginia, 11, 140, 144, 154
Virginia case study, 91–92
  clearance time prediction, 112–20

data analysis, 98–112
incident duration, 121–23, 232
    resource analysis, 154–55
    resource allocation, 155–60
response plan, 144–45, 232–33
results comparison, 123–25
    rule base response, 160–62
survey forms/data collection, 92–98, 154
Virginia Department of Transportation,
    93, 95, 142–45, 154, 167,
    170, 233
Virginia State Police, 93, 96, 145, 154
Virginia Tech Center for Transportation
    Research. *See* Virginia case study
VMS. *See* Variable message sign
VMT. *See* Vehicle miles traveled

WAIMSS. *See* Wide-area incident management support system
Washington, D.C. congestion, 3, 6

Weather conditions, 148, 179, 192
Weather-related incident, 112
Wide-area incident management support
    system, 17, 41–42, 129, 136
application design, 45–50
conclusions regarding, 231–34
decision support modules, 46–47
delay calculation module, 48–49
diversion system architecture, 172–77
duration estimation module, 47–48
framework for integration, 43–45
overall concept, 42–43
response module, 49–50
software implementation, 50–57
Virginia case study, 136–37, 154–62
Work force demographics, 2
Wrecker involvement, 109–11, 123, 156

YAD. *See* Young/aged driver
Young/aged driver, 63

Recent Titles in the Artech House ITS Library

John Walker and Chelsea White, *Series Editors*

Advances in Mobile Information Systems, John Walker, editor

Incident Management in Intelligent Transportation Systems, Kaan Ozbay, Pushkin Kachroo

Intelligent Transportation Systems Architectures, Bob McQueen and Judy McQueen

Positioning Systems in Intelligent Transportation Systems, Chris Drane and Chris Rizos

Smart Highways, Smart Cars, Richard Whelan

Transport in Europe, Christian Gerondeau

Vehicle Location and Navigation Systems, Yilin Zhao

Wireless Communications for Intelligent Transportation Systems, Scott D. Elliott, Daniel J. Dailey

For further information on these and other Artech House titles, including previously considered out-of-print books now available through our In-Print-Forever® (IPF®) program, contact:

Artech House
685 Canton Street
Norwood, MA 02062
Phone: 781-769-9750
Fax: 781-769-6334
e-mail: artech@artech-house.com

Artech House
46 Gillingham Street
London SW1V 1AH UK
Phone: +44 (0)171-973-8077
Fax: +44 (0)171-630-0166
e-mail: artech-uk@artech-house.com

Find us on the World Wide Web at:
www.artechhouse.com